DATE DUE

GAYLORD | | | PRINTED IN U.S.A.

MANY
SKIES

MANY SKIES

Alternative Histories of the Sun, Moon, Planets, and Stars

ARTHUR UPGREN

RUTGERS UNIVERSITY PRESS
NEW BRUNSWICK, NEW JERSEY,
AND LONDON

Library of Congress Cataloging-in-Publication Data

Upgren, Arthur R.
Many skies : alternative histories of the sun, moon, planets, and stars /
Arthur Upgren.
p. cm.
Includes bibliographical references and index.
ISBN 0-8135-3512-3 (hardcover : alk. paper)
1. Solar system—Miscellanea. 2. Astronomy—Miscellanea. I. Title.
QB502.U64 2005
523.2—dc22
2004007533

A British Cataloging-in-Publication record for this book is available
from the British Library.

Manufactured in the United States of America

TO BART BOK AND ARTHUR BEER

With gratitude

CONTENTS

CONTENTS

MANY
SKIES

Blessed is he who learns how to engage in inquiry,
with no impulse to harm his countrymen or to pursue
wrongful actions, but perceives the order of immortal
and ageless nature, how it is structured.

EURIPIDES

INTRODUCTION

Nothing could be more obvious than that the Earth is stable
and unmoving, and that we are in the center of the Universe.
Modern Western science takes its beginning from the denial
of this common-sense axiom.

DANIEL J. BOORSTIN, *THE DISCOVERERS*

The sky we see is a familiar one, with few changes taking place over the
course of a single human lifetime. Our current view results from a delicate balance between many factors, and changes in just a few of them
would have given us a sky with a very different appearance. For example, what if we had more than one sun in the sky? Or more than one
moon? In either case we would see a panorama transformed. Alternative history, in which the consequences of various "what ifs" are explored, has become very popular in recent years, at least within the
realm of human history. But of greater merit is the comprehension it
can provide as to the ways in which ours is but one of many worldviews
based on the world we experience. What if the South had won the Civil
War; what if Napoleon Bonaparte had won the Battle of Waterloo? Or
what if the Constitutional Convention that met in Philadelphia in 1787
had deadlocked over the issue of slavery and so adjourned without
further progress? Any of these alternatives would metamorphose the
world we know into something else. So it can be with astronomy.

In this book we examine a number of alternative solar, stellar, and
galactic systems and arrangements, and follow their consequences not
only on the sky we would see, but also on the variant historical interpretations that might have come to pass, as the science and discoveries
of astronomy might occur much earlier or later than they did. If Aristotle had detected a parallactic shift in the motions of stars reflecting
our planet's motion about the Sun, for example, would it have led him

to the Sun-centered cosmos nineteen centuries before Copernicus proposed it? History might have been very different, not only in science but also in other fields of human endeavor. These may include religion, exploration, the formation of nations and empires, and changes in literature and the arts.

Until the invention of the telescope in 1608, only two among all the celestial objects could be seen as other than points of light: the Sun and the Moon. One of them circles around us and we orbit around the other in tight, nearly circular paths; as a consequence, both retain a nearly constant angular size in the sky. If even one more object were visible as a disk to the naked eye, whether moon, planet, or star, our world would be very different. Any such disk would be close enough to be among those that are seen to move over days, months, or years and would be likely (although not with certainty) to change its angular size over time. Astronomical progress throughout history would likely have been more rapid, with advances occurring in a compressed time frame over the twenty-five centuries since the Hellenic natural philosophers first brought to rational thought the concept that no division exists between *physici* and *theoretici;* that is, between that which is or may someday be explained in terms of natural phenomena and that which must be left to the divine and miraculous, created by deities. All things are eventually explainable, the ancient Greeks surmised, a unique step in the development of our species, and some conjectures of what might have been are presented here.

The Greeks had one other new development going for them—this was the alphabet, the system of letters representing vowel and consonant sounds that had just replaced hieroglyphics, the earlier languages that used word pictures to describe thoughts, moods, and concepts as best they could. The subtlety of expression available with the alphabet made for radical gains in the descriptions of concepts that formed so much of Hellenic thought.

I know of only one other book that explores the alternative history theme in astronomy: *What If the Moon Didn't Exist?* by Neil F. Comins, published in 1993. Comins limits the consequences he examines largely to the resulting changes in the overall history and evolution of the Earth and the life forms that have evolved upon it. We choose here to emphasize the altered appearance of the sky, particularly the night sky, and the major attempts to interpret the observations made of it.

I have not, in this book, calculated with great precision many of the properties that an altered solar system or universe might entail. The data are correct within limits that adequately convey the appearance of the sky with the alterations made. This choice devolves in part from the inevitable limitation in the precision of the observational data for the stellar and galactic realms. Within the solar system we can apply values for distances, masses, and the like, to an extraordinary degree of accuracy. But once we pass on outward into the much more distant stellar system, we often contend with one- to two-place levels of accuracy in these same features for the stars and beyond. I have also not changed the laws of physics that apply everywhere; E still equals mc^2, and force still amounts to mass times acceleration. A change in either of these would impose such profound alterations of the universe that those I do make seem petty in comparison. If gravitation varied not as the inverse square of the distance but as its inverse cube, the cosmos would be scrambled beyond all recognition and the differences depicted here would become meaningless.

Moreover, it is easy to become ensnared in the true meaning of commonly used language. On the witness stand under oath, questions such as "Does the Earth go around the Sun?" or "Do planets move in elliptical orbits?" must be answered in the negative. In the full exploitation of the meaning of Newtonian mechanics, particularly as emended by Einstein, we know that the implied simplicity of questions like these covers a multitude of exceptions. I have chosen here to continue to emphasize the contrast between the earlier answers to the above and those first gained during and just after the Copernican revolution; thus, the Earth does go around the Sun, which does not go around the Earth, and planets move much more closely to elliptical paths as opposed to the circular orbits fashionable in the older models. These remain at the levels of appreciation of the generally educated reader. Had we only the stars and galaxies upon which to establish a mechanical system, we might know no better than the more simplistic (purely elliptical) form for an orbit, since the nuances arising from perturbations of additional nearby objects and relativistic motion might not yet be detectable. We would be in danger of promoting a tautology that we in fact know need not exist.

We know, too, that many of the anomalies inside and outside the solar system, such as the freak encounter that formed the Moon, or the wild tilt of the axis of rotation of Uranus, are likely examples of non-equilibrium statistical physics. Related to chaos theory, this nascent discipline appears to account for many human and natural phenomena of varying sizes. Earthquakes, mass extinctions, stock market levels, and even wars follow a power law, which strongly implies that the large and rare catastrophic occurrences in each category are simply the cases in which small events give rise to an avalanche of much larger proportions for reasons not entirely clear. One more grain of sand placed on a sand pile, especially if the pile is critically near the angle of repose, has a chance to collapse much of it into a lower, more disordered heap. This forms a further indication that a rerun of the many small events that occurred in the past would have led to a very different bunch of planets even if we could repeat with all possible precision the conditions of their formation. If their formation processes were started over again, the solar system and the galaxy would not have formed in the way they did, any more than human, weather, or geologic history would have done. This extensive new field of historical physics puts the lie to those who maintain that science is at or near an end, with only a few more details to further understand and explain.

Alternative history in any field can reveal the conditions required for its constraints to be allowed to vary, or to remain constant. With even the slightest alteration of the historical scene, Alexander or Napoleon, Newton or Shakespeare might never have been conceived and our present civilization would be markedly different. So in science—if blobs condensing in orbits around the Sun had been of slightly different mass than they were, even Genesis might well have no relation to the one we know.

No one, not even many ophthalmologists, has the experience with such faint sources of light as do astronomers in the course of their normal observations. Think of a bead of dew atop a blade of grass reflecting and focusing the beams from a distant streetlight a block or more away. The tiny point of light from the dewdrop is probably many times brighter than most of the stars we see on a clear night. A normal candle seen a mile away outshines our first-magnitude stars, and at five miles it would still be visible to the naked eye. In perhaps no other activity do scientists observe and measure with care and precision points of light as faint as these. Light sources shining at only a billionth of the light of the faintest star visible to the eye can now be viewed and recorded by our largest telescopes.

The study of the sky, whether dark or light-polluted, is the astronomer's province and his or her laboratory—for this reason we seek to protect our laboratory from degradation from any source, for astronomers and for all to see this most universal and sublime of spectacles.

I
THE SUN AND
THE MOON

1
OUR THREE
MOONS

Our Moon, our eternal companion, our queen of the night, was the result of an accident. The Earth has been accompanied by its faithful moon almost since the formation of the solar system 4.6 billion years ago. We know that there was much more material flying around in those early days of the system than there is now; planet-sized chunks not ready to settle down into final, proper nearly round orbits revolving about the star in the middle of it all. Seemingly, these renegades flew about like loose cannons, but in fact, they, too, faithfully followed the laws of motion and gravitation.

Proto-Earth, proto-Venus, and proto-Mars, our names for these three planets during their coalescence, were then very busy condensing and solidifying into the dense planets of today with material like silicon and iron and oxygen making up the bulk of their interiors. But nature was not done with proto-Earth, not at all. Its biggest calamity came early on in its history like a charging rhinoceros onto a crowd of picnickers and hurtled into our nascent planet head-on. A Mars-sized chunk piled into the Earth and knocked a swath of stony silicate stuff out and away from our globe, but not so far as to escape its gravity altogether. No, the chunks torn out of our midriff fused back together just out of reach and there they coalesced to form the Moon, our big satellite that makes of us almost a double planet.

Until Pluto's moon, Charon, was discovered in 1978, we appeared the closest thing to a double planet, a barless dumbbell, among all the planets. The Moon, 240,000 miles off, with a quarter of our diameter (2,160 miles) and 1/81 of our total mass, gives us a night sky with something special in it. By far the most gazed-upon object by *Homo sapiens*, visible to people in every land, the Moon graces our night as does nothing else. Poor Mercury and Venus have no moons to accompany them on their rounds, whereas we have a big one. Mars has two but they are tiny potatoes that, like as not, were asteroids that strayed too close to that planet and were captured into orbit about it. Their shapes are not globular because, being only 10 to 15 miles across, their gravitational fields are too weak to overcome the natural rigidity of their stony structures.

The giant planets, those great gasbags of the outer system—Jupiter, Saturn, Uranus, and Neptune—have dozens of moons apiece; we don't know how many. At latest count Jupiter has a whopping 61 moons; Saturn has 31; Uranus, 27; and Neptune, 13. Pluto and even a handful of small asteroids each have a moon. We have found no moon that has its own moon, but this is not impossible.

In this chapter we imagine that our world has not one, but three natural satellites, each one named for one or another moon goddess. We would not use the expression "the Moon" if we had more than one. The largest and most distant of the triad we shall call Selene after the Greek name for the moon goddess, and it is identical in all respects to the moon we have; Artemis is smaller and closer, and Astarte is tiny and yet closer, appearing only as a point of light in our imaginary sky. These, then, lead to the alternative case of this chapter's scenario.

In the deepening evening twilight the fireflies could be seen winking at each other through the trees. The summer had proceeded past the high point, after which the nocturnal silence was replaced by the melismatic chirping of legions of insects persisting throughout the warm night. Far aloft between the branches, the stars were also making their steady appearance in the crepuscular gloom, the brighter ones first and the others filling in the celestial gaps between them. Selene shone brightly at the first quarter phase high in the southern sky. A brightening lambent glow in the east heralded the rising of Artemis. Although a third smaller in angular appearance than Selene, it was far brighter because it was in its full phase.

Selene shines with the silvery color so often given to it through song and legend, but it contrasts to its detriment with one of the newer types of white metal halide street lamps visible down the quiet, leafy suburban street. The lamp defines white as Selene never will, revealing the satellite to show a slightly dull yellow-gray aspect. As full Artemis rose above a distant line of trees, we could sense its slight reddish tinge, just noticeable when contrasted against Selene's color, although when seen alone in the sky Artemis seemed as colorless as Selene. We glanced at tiny Astarte, the closest and much the smallest moon rising a little ahead of Artemis and appearing as a bright star, being too small to subtend a noticeable disk in the sky to our unaided eyes.

A cloud here and there could be seen around and through the foliage. The clouds were small and of the puffy cumulus type that frequently dissipate before sunset, when sunlight no longer heats the ground and their thermal supply is terminated. But sometimes they can hang on longer, as these have done. Their fringes were limned with moonlight on all sides from one moon or the other, but unevenly. Those closer to Selene in the sky were dominated by its light, and the ones to the east near Artemis reflected its moonbeams more vigorously. A contrail all but touched the disk of Artemis and glowed brilliantly at its tangent point to that moon. Moondogs appeared on either side of the just risen disk, revealing ice crystals in the upper atmosphere.

In the vignette above, we stated that Artemis at full is much brighter than Selene at the first (or last) quarter phase, when it appears as a half-moon, because at the quarter phase a moon gleams with a considerably lower luminosity than it does at full. This results from two separate effects. First, in much the same way that a northern exposure retains snow cover longer than its southern counterpart (due to the oblique angle with which the Sun's rays strike it), the quarter phase shows us a surface receiving sunlight from the side at a substantial angle, not full on. Second, the face of any moon is pitted with countless craters, mountains, and rills. From the side we see much of its face in shadow whereas from full on we see no shadowy contrast at all. At the full phase a moon is no more revealing of its surface than the white English subtitles in a movie I once saw that featured them against the white buildings on the sunny Cyclades islands in the Aegean Sea. Taken together the two effects raise a moon at the full phase not to twice, but to fully 10 times the brightness at the quarter phase of the same moon.

Both of the larger moons display blotched faces, scarred by billions of years of impacting meteoroids large and small. Selene's puffy, off-center man-in-the-moon visage and the pointed aspect of the face on the smaller satellite, resembling no creature more than a weasel, are well known, as they both keep the same side toward the Earth, locked by the planet's tides into a rigid embrace. Each of the three moons reflects less than 10 percent of the incident solar illumination received. This reflecting power, called albedo, is always low for rocky, airless worlds such as themselves and Mercury, which exhibit only a tiny trace of air of any kind. Mars has a thin but distinct atmosphere, Earth a much thicker one, with a surface half obscured by clouds at any one time, and Venus and the giant planets have atmospheres that are perpetually cloudy and forever concealing any hint of a surface lying underneath. Consequently, the albedos of Mars and the Earth are near 15 and 35 percent, respectively, and those of the cloudy worlds are near 60 percent, clouds being superb reflectors of sunlight.

Artemis is only about 700 miles in diameter, about a third Selene's size but, being only half as far away, it appears two-thirds the angular size of Selene in the sky as seen from the Earth. Artemis is too small to totally eclipse the larger disks of the Sun and Selene, although annular eclipses are common—those semispectacles in which a ring or annulus of the larger disk remains visible surrounding the smaller, closer one. Artemis moves faster, taking not quite 11 days to circle the Earth, unlike the 27$\frac{1}{3}$-day true period of our largest moon, or a 29$\frac{1}{2}$-day lunation period (due to our swing around the Sun each month, it takes over 2 days longer for Selene to repeat a phase than it does for it to return to a particular point in the sky). Whenever the two moons are near each other in the sky they show the same phase to us because the phase depends only on the angle between a moon and the Sun as seen from the Earth.

To complete the Earth's family of natural satellites, Astarte, the smallest and closest moon, appears a pointlike object to the unaided eye but its phases can be seen with binoculars. Only 15 miles across, it is seen through the telescope as a lumpy potato in appearance, not having the necessary gravitational field to force it into a glove of rondure like its much larger neighboring moons. It is probably a captured asteroid, as are its near twins in size, Phobos and Deimos, the two small moons that orbit Mars, and the fleets of small moons accompanying each of the outer giant planets. This evening Astarte is nearly full and shines as a starlike point of light of magnitude -2, outshining every star in the nocturnal sky, almost as bright as Jupiter appears to us, and of about the same color as that yellowed globe.

Before the space programs and the subsequent landings on all three moons, it was thought that they were composed of the same combination of silicates and other minerals, but now we know that mass promotes a difference in stratification of the interiors of objects of different sizes. None are known to have an atmosphere or standing water, essentials for life to form, and none have any source of internal heat, necessary for volcanism and other seismic activity.

Although each moon moves along its own unique orbital track in the heavens, they all move nearly in the same plane and do occasionally pass in front or behind each other. We have transits upon occasion of

Astarte passing across the disks of either of its larger neighbors, Selene and Artemis, and all pass in front of the Sun from time to time. The celestial mechanics of the system are almost impossible to calculate, but computational capabilities are now equal to the task. We know, for example, that in 883 B.C. Astarte passed across the disk of Artemis just as the latter was passing in front of Selene, although no record of any observation has been found. Only Homer, among famous names, lived near this time and he makes no known mention of the event. This is the only close syzygy, or lineup, involving the Earth and all three of its companions in recorded history, and it will not happen again until the year 3249. One of the most arresting spectacles among all of the sky's wonders is the sight of Artemis passing close enough to Selene to cover a fair portion of its face. At that moment the pair appears to separate, and they resemble nothing so much as an amoeba just at the moment of division into two. Astarte, like Artemis, is of a slightly reddish tint compared to the silvery Selene or the yellow-white Sun. This is due to small deposits of iron oxide on their surfaces, the same rusty substance that gives Mars its ruddy hue; they probably got slightly coated with the stuff during the early formative days of the Earth system, some 4.6 billion years in the past. The seismically active surface and wind erosion of our large world guarantees that any trace of the red material here vanished eons ago. Normally, three bodies of the mass and such close proximity as the Earth, Artemis, and Selene make for an unstable, disruptive situation in which one of the three is ejected altogether, but here the system appears to be stable, at least for the present.

The mass of a person standing on the surface of the Earth is directly proportional to and in fact is defined as his or her weight. But elsewhere the weight depends directly on the mass ratio of another body to the Earth and is inversely proportional to the square of its radius compared to that of the Earth. The actual equation, as derived by Isaac Newton, is given as:

$$F = Gm(1)m(2)/d^2$$

where F is the force of gravity, G is the gravitational constant, m are the masses of the two bodies, and d is the distance between their centers.

Thus a 180-pound man would weigh only one-sixth as much or 30 pounds on Selene, and only 9 pounds on Artemis. On tiny Astarte, the same person would tip the scales at less than a meager 0.01 ounce (5 grams); in fact, anyone standing on Astarte with a good arm could throw a baseball free of this minuscule world and into its own orbit about the Earth. The total surface area of Selene is around 14 million square miles, a bit larger than that of Africa together with the Arabian Peninsula. The surface of Artemis measures only 1½ million square miles, comparable in land area to about half the forty-eight contiguous states of the United States of America. In sharp contrast, Astarte comes up with but 800 square miles of surface area, smaller than that of Luxembourg or Rhode Island. Since the time of Pythagoras, five centuries before the Common Era, the correct distance order of the three satellites from the Earth has been known. Plato (428–347 B.C.), Aristotle (384–322 B.C.), Eudoxus (409–356 B.C.), and the other astronomers of the Athenian School correctly placed Astarte in the lowest orbit just above the mundane earthly domain and its four elements, yet firmly within the heavenly realm as the nearest bit of their divinely pure quintessence. This starlike object requires just under 7 days to make one trip about us, compared to nearly 11 days for Artemis and 27⅓ for Selene. The three are followed in Aristotelian distance order by Mercury, Venus, Sun, Mars, Jupiter, and finally Saturn, just this side of the orbit assigned collectively to the stars. These are the nine bright objects seen to move among the network of constellations we call the zodiac, and each gives its name to one of the 9 days of our week. Although one day is set aside and named for each moon, each of the five bright naked-eye planets, and the Sun, their scrambled order speaks to an epoch well before that of classical Greece, when for the first time scientific principles were used to study the cosmos and an internally consistent if not correct distance order was established.

Our three moons present an enormous variety of phases and brightnesses, and the night sky is an ever changing pageant of beauty. Every astronomer knows that whenever the two larger moons appear in the crescent phase together near Venus as an evening star in the western sky after sunset, the telephone goes crazy. Two thin crescents alongside the brightest planet close together in the sky fire up the public's attention as does no other celestial arrangement except a total solar eclipse by Selene, when the ghostly faint solar corona can be seen. Even in the crescent phase Astarte is among the brightest stars in the sky and only adds to the spectacle if it too is nearby. Our triple-moon system would above all provide further interest and color to an already rich night sky. And when either of the two larger satellites appears near the full phase, the songwriters get busy rhyming their moons, Junes, and lagoons.

Sometimes we see the entire disk of each of the larger moons faintly when it is in the crescent phase (unlike the moon in Hollywood horror films, which is always shown as full). As Leonardo da Vinci was the first to explain, the illumination of the entire lunar disk in the crescent phase is due to the pale ashen light, called earthshine, reflected from a nearly full Earth in the opposite phase from that of the Moon at the time. The Earth as viewed from Selene is about 40 times as bright as that moon seen from here in the equivalent phase because the Earth is almost 4 times the diameter and its albedo is several times that of the airless Selene.

One new feature would be found in this three-ring circus we have imagined. Just as Selene can enter the shadow of the Earth, providing us with a lunar eclipse, so could both of the other smaller moons enter the shadow not just of the Earth, but of Selene as well. In the Earth's shadow, a moon glows with a dull coppery hue, due to the fact that our atmosphere refracts a small bit of sunshine into the shadow. Artemis could just fit inside Selene's shadow, and in this case we could watch it disappear altogether, becoming totally invisible, since Selene has no atmosphere to refract a little copper-colored sunlight into its shadow, as the Earth does. But imagine the situation with the larger two moons. Selene is in the thin crescent phase and if Artemis passes directly behind it as seen from the Sun, that smaller moon will plunge from its luminous, sunlit crescent phase into the darkness of Selene's shadow; it will darken but not quite to total invisibility, for now it is illuminated by earthshine and its full disk appears to us with that faint ashen gray earthshine. For a few minutes this effect would continue, and then Artemis passes back into full sunlight. The ethereal beauty of this spectacle would be unique among the planets of the inner solar system.

The two satellites visible as lunar disks do not have much chance to enter the shadow of the Earth, thus bringing about a lunar eclipse, at the same time. Their simultaneous appearance with the particular copper-colored hue that moons have in our shadow is rare indeed and has never been observed. This is because their orbits, like any others, form great circles in the sky, great circles being the largest possible circles projected onto a sphere. Any two great circles intersect at two and only two points that always occur opposite each other in the sky; these two points, called nodes, are the two equinoxes in the case of the intersection of the celestial equator with the ecliptic. The nodes are rarely stationary but move slowly among the stars; precession carries the equinoxes around the sky in the retrograde direction once every 25,800 years. Similarly, the Moon's orbit, inclined about 5.2 degrees to the ecliptic, rotates about it such that the nodes circle the sky in

18.6 years in a motion called the nutation. Only at or near the nodes is a lunar or a solar eclipse possible; otherwise Selene would miss the Earth's shadow as it does most times. Artemis would lie in a different orbital plane and thus have nodes of its own; only if they are coincidentally very close to the nodes of the orbit of Selene around the Earth and the ecliptic could the two be eclipsed at the same time, and that would be rare indeed.

Although Artemis is a trifle less dense than Selene, and only one-thirtieth its mass, the tides it raises are fully one-fourth those raised by the larger moon and almost three-fifths the tides due to the Sun. This is because the tide-raising power of one object on another is approximately dependent on the inverse cube, not just the inverse square, of the distance between them. Even with Astarte's puny effort omitted from consideration, the tide tables are the very devil to compute. Astarte's tidal force is only about 1/70,000 that of the Sun, and less than those imposed by Venus and Jupiter when they are nearest the Earth. Although all bodies in the universe affect all others tidally, any but the largest and closest can be ignored for all but the most precise calculations.

The tides on the Earth and its oceans go through their largest amplitudes or spring tides whenever the larger two moons are both at the new or at the full phase, or one moon is at each of the two phases—then they and the Sun are all in a line and they all pull together in the same direction, their tide-raising powers reinforcing each other. With Selene at the quarter phase and Artemis at new or full, pulling along with the Sun, we experience the most even balance between tidal forces and hence our smallest tidal variations, the so-called neap tides.

Selene, our real Moon, accounts for 2.2 times the tides raised by the Sun, but that is only an average. Both move in elliptical orbits, thus the Moon passes from perigee, its closest point to the Earth and where its tidal pull is greatest, to apogee, its farthest, and the point of minimal effect. The Earth passes through perihelion and aphelion, the names for the nearest and farthest points in its orbit with the Sun. Whenever the Moon is at perigee and the Sun is at aphelion, the ratio increases from 2.2 to nearly 2.8 to 1.0, and at the reverse conditions, just 1.7 to 1.0. Furthermore, the latitude on the Earth and the positions of the Sun and Moon impose additional variations on this ratio. When we add another moon to the picture we shall only deal with average circumstances—we do well to leave such complexity alone.

On average, our one Moon produces spring tides that swing through about 2.5 times the amplitude shown by the neap tides, although the irregularities of coastlines can often make this difficult to observe. Spring tides occur at new and again at full Moon, when it is in line with the Sun and the two reinforce each other. At first and last quarter, the Moon counteracts the weaker Sun and the neap tide condition appears. During the spring tidal times, when tides swing through their greatest amplitudes, water in the Bay of Fundy between New Brunswick and Nova Scotia manages to reach a high tide some 50 feet higher than the low tide just 6¼ hours later, the greatest swing observed anywhere. Yet a week later at neap tide the amplitude is less than half as great.

When we add another substantial mass in the form of a second moon to the arrangement, the complexity becomes evident. We require all of the players to line up in order for the tidal effect to be maximized. At some point the Sun, Selene, Artemis, and the Earth must all be in a line or very nearly so. Then all pull together and the gravitational tug on our oceans swings through the greatest amplitude from the highest of high tides to the lowest of low tides a quarter of a day later. But in this case it is not a simple matter of waiting a week until the Moon (Selene) and the Sun are at quadrature; that is, at right angles to each other, for tides to be at their weakest, because in that interval, Artemis will have moved to an intermediate position. Since the latter moves with an 11-day period, it will be between full and last quarter when Selene reaches first quarter, whenever both start out at new. We must wait for two more weeks until Selene reaches the third quarter and Artemis has circled the Earth just about twice before a lineup is once again achieved. Only then will conditions be ripe for a neap tide to occur.

At the Bay of Fundy the spring tide may reach an amplitude of 70 feet under these two-satellite conditions. Then, three weeks later when the next neap tide appears, the swing may only be about 10 feet. Finally, after another three weeks have passed, some 22 days later, after two complete orbits of Artemis and three-quarters of one by Selene, all reach their proper stations for the next spring tide to take place.

There comes a point when the two larger moons appear equally bright. With similar albedos, the point will occur whenever the smaller moon is somewhat closer to the full phase than is the larger satellite. The phase angle is defined as the angle between the Sun and the observer on Earth as seen from the moon in question. Thus at full the phase angle is zero and at the quarter it is 90 degrees. The object's brightness is a factor of the so-called Bond albedo, or A, the total percentage of sunlight reflected in all directions. This is the product of two other factors, the geometric albedo, p, the percentage of light reflected at a phase angle of zero (full moon), and q, the phase integral determined by observation; thus $A = pq$. Moonlight varies widely with phase angle. It was noted above that a full moon of surface properties such as are found on Selene and Artemis is about 10 times as bright as the same moon at the quarter. With similar albedos, the smaller moon shines with 4/9, or just under half the total brightness of Selene. Thus whenever Artemis is nearly 30 degrees closer than Selene to the point in the sky opposite the Sun, the two shine with the same amount of light.

One final note on the moon business. The orbit of our real Moon—Selene in this portrayal—is not static. The tidal force of the Moon on the Earth acts to slow down the rate of rotation, our day. It does so by minuscule amounts but over the vastness of time they have a way of accumulating to a very large amount. Early in our planet's history the day was but 8 hours in length, but with the perpetual tugging of the Moon on our oceans, the day lengthened; at present it lengthens by 1/50,000 of a second every year, accumulating to the point that we must intercalate an additional second every several years in order for the slower rotation to catch up to our atomic clocks, which now define time for us with total precision. Once every few years, the ball overlooking Times Square in New York must register 61 seconds in its fall, not 60 seconds in the minute of its descent. Thus on December 31 of these special years, the second after 11:59:59 P.M., or 23:59:59, is not 12:00:00 midnight or 0:00:00 by the 24-hour clock, but instead reads 11:59:60 or 23:59:60 before going on a second later to midnight and the new year. The old year is 31,536,001 seconds in duration, not 31,536,000 (with another 86,400 seconds for leap years in both cases). Over historical times the total deficit runs up to a few hours, enough to displace the path of an ancient total solar eclipse from its calculated track using constant 24-hour days to points up to several hundred miles away.

The year lengthens as well, but by only 3 minutes every billion years, a negligible amount by any measure of historical reckoning. Even as it stays the same length, it will include fewer days as the day itself lengthens. There is a quantity in physics called angular momentum, which in any system *must* be conserved. Angular momentum is a name for the total amount of spin of all masses in the system, by rotation and by orbital revolution. In slowing down our day the Earth loses angular momentum, which must be gained elsewhere, in this case by the Moon. It requires our satellite to move away from the Earth by about 5.8 centimeters per year, lengthening the lunar month ever so little. Atomic clocks keep time much better than the Earth, Moon, and Sun, and we use them to keep track of and now to define the correct time. After billions of years, long after the Sun has evolved into a giant and fried the closer planets, the Moon would be about twice its present distance from us and take some 43 days to circle the planet. At this point we would keep the same face toward the Moon as it already does to us. A barless dumbbell would result at that belated epoch, and the Sun would rise and set on both worlds but once in about 43 days.

I have ignored the angular momentum considerations in the three-moon scenario—I am aware that it would interject itself into this more complex system, but I hesitate to take the time to figure it all out.

2
WITHIN A
TRIPLE STAR

We assume here that the Sun is one of three stars in a system that closely resembles Alpha Centauri. We can still refer to it as the solar system, since our Sun is the largest, closest, and brightest of the lot. At a distance of only 4 light years, Alpha Centauri, seen in the deep southern sky as a single star, is the nearest of all stars and the third brightest in all the heavens, outshone only by Sirius and Canopus. But it really consists of the three stars that are closest to our solar system. The brightest of the three, Alpha Centauri A, is a nearly exact twin of the Sun in size, magnitude, and color and which will be known in our story as Helios, the Greek name for the Sun itself. It can be considered to be the Sun since it is in every way almost identical to our Sun. Then in addition we have twins of Alpha Centauri B and C in our system, both smaller and fainter than A (or Helios), C being much farther from Helios than B. Alpha Centauri C is also known as Proxima because it lies slightly closer to us than the brighter pair (4.22 as against 4.35 light years). Here B is called Osiris after the Egyptian sun god at the moment of dusk when its light is subdued, and Proxima retains its name.

As Helios disappeared below the horizon, the sky was clear and our shadows were still visible and quite short since Osiris remained high overhead. Helios slipped below the horizon and we entered the point of equivalence, that time in the evening when light from Helios dimmed to the level of the light from Osiris. The crescent Moon shone mainly by the yellow-white light of Helios, but a fainter orange crescent on the other side of the Moon lit by Osiris could already be seen.

This lovely spring evening promised to be very clear, dark enough to spot the most distant member of our solar system, Proxima. The faint star is just nicely visible to the naked eye in the constellation of Libra, and we will look for it later when it gets fully dark.

The ancient Greeks appear to be the first to codify celestial observations into a scheme or paradigm. They sought to preserve the phenomena; that is, to require theory not to conflict with observations. In order to speculate, as we will, on the science they would have produced under conditions described in coming chapters, we need to be familiar with the successes and failures they did have in order to appreciate the changes their perceptions might take under these alternative conditions.

These people were not topflight observers. Many other societies, such as the Babylonians and the Mayans, made far more precise and consistent observations and records of the phenomena in the sky. But in saving the phenomena the Greeks became scientists as no others would for centuries. In the sixth century B.C. it began, as best we can deduce, with Thales, Pythagoras, and others of the Ionian school on the shores of Asia Minor, who first became aware that the Earth is round and also of the nature of lunar and solar eclipses. The Athenians, particularly Plato, Aristotle, and the more mathematically inclined Eudoxus, developed a scheme with variations that held sway in some form until the sixteenth century after Copernicus published his counterplan.

In the late fourth century B.C. Alexander conquered much of the known world and founded his empire, centered on the new city of Alexandria at the mouth of the Nile delta, the largest of the many cities and villages named for him. The city quickly became the learning center of the region and the world. The astronomers who lived and worked there far surpassed their predecessors in scientific achievements and brought to those Hellenistic times an awareness and richness of analytical detail unequalled until the Renaissance.

One of the greatest of the Alexandrian astronomers was Aristarchus, originally of the Aegean island of Samos. He lived from about 310 to 230 B.C. and used algebra and trigonometry to derive the distance to the Moon and Sun. The Moon, alone of all celestial objects, can be seen to reflect the Earth's rotation in a single night. From this, Aristarchus found that the Moon is about 30 Earth diameters away. He used this and observations of eclipses to find the size and distance of the Moon. Then in a perfectly sound and logical way he deduced that the Sun is 19 times as far as the Moon. He did this in two ways—he used the data produced by lunar eclipses, but he also measured the angle between the Moon and the Sun in the sky at the moment the Moon appeared exactly in the half phase.

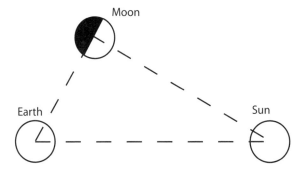

Fig. 2.1 The method used by Aristarchus to determine the distance to the Sun in terms of that to the Moon, a quantity he knew. The angle at the Moon is a right angle. The angle at the Earth is a function of the ratio Earth-Sun/Earth-Moon distances.

The line separating the bright and dark sides of the Moon is called the terminator. This line would be straight at the first and last quarter phases (see Fig. 2.1). Aristarchus deduced that the triangle formed by the Sun, Moon, and Earth at the moment when the terminator—the line across the lunar surface that divides sunlight from the dark side of the Moon—is a straight line, making two equal hemispheres as seen from the Earth. At this moment, the angle not at the Earth but at the Moon is 90 degrees, a right angle. Since all three angles of a plane triangle always add up to 180 degrees, a measure of the angle between Sun and Moon will be just a little less than a right angle, and that at the distant Sun will be very small. This method is sound but the angle is very difficult to ascertain with the naked eye. Aristarchus found it to be 87 degrees and concluded that the Sun is about 19 times as remote as the Moon; thus, it must be 5 times the diameter of the Earth. This turned out to be the first occasion on record where it was determined that something up there in the sky was larger than our big, seemingly flat world. The modern size of the angle is 89°50', just 10 arc minutes short of a right angle, placing the Sun at almost 400 times the Moon in distance and thus nearly 400 times the size of the Moon and 110 times that of the Earth. Aristarchus knew that the Moon's distance was about 60 times the radius of the Earth; therefore he could convert relative sizes and distances into absolute ones, albeit with considerable error.

When Alexander returned from the conquest of Persia and other lands to the east, he brought with him the records of observations by the

Babylonians, which were far superior to those available to the Greeks of the time. The frequency of clear weather and the clarity of good nights helped to inspire their finer observing tradition, but they also seemed to make observations for their own sake (as many others have done since) rather than to be compared to results predicted from theory.

After the Macedonian conqueror founded his great namesake city, learning accumulated there, and Alexandria rapidly became the primary site of intellectual effort. The great Alexandrian library became an amazing repository of perhaps a million works of the Hellenic and Hellenistic greats in all fields, to which Roman works were added later. With the better observations in hand, astronomers there were led to an improvement in theory. Where Eudoxus had rings and invisible crystalline spheres, one for each separate motion undergone by the Sun and Moon and each planet, Aristarchus came up with something new. He proposed that the Sun, not the Earth, was the true center of the solar system; this 1800 years before Copernicus proposed the same thing.

Copernicus appears to have been aware of Aristarchus's heliocentric model when he made up his own (this point is discussed in detail by Owen Gingerich in *The Eye of Heaven*). What happened to obscure the early Sun-centered triumph? Several factors are cited to explain this lacuna in science. First, the destruction of the great Alexandrian library by Christians and later by Muslims eliminated records of the writings of Aristarchus himself on the subject. Second, as Thomas S. Kuhn cites in *The Structure of Scientific Revolutions,* the world was not ready to make of the discovery a new paradigm for other scientists to build upon. Aristarchus appears to have had no followers and his heliocentric system was soon discarded by all. His later followers, the three best known being Eratosthenes, Hipparchus, and Ptolemy, all returned to the geocentric layout since the alternative theory did not have enough observational support to convince even these illustrious astronomers of its worth. Kuhn compares this paradigmatic change at the time when Copernicus reintroduced heliocentrism in 1543 to later ones and notes that even in the sixteenth century it encountered tough sledding. Kuhn likens it to two more recent scientific revolutions—the one in chemistry in the 1770s, when Antoine-Laurent Lavoisier and others rejected the concept of phlogiston in favor of oxygen and its reactions. The other is the relativistic revolution by Albert Einstein in 1905, when he published his special theory of relativity in response to perceived failures in Maxwell's theories and the concept of the aether of the 1880s and 1890s.

Both of these examples took far less time for the new theory to re-place the old than did heliocentricity over geocentricity after 1543. This is because extrascientific influences, such as theology and scripture, interfered with the acceptance of the Sun at the center. Pressure came from the Vatican and also, more strongly but with less impact, from the new reformed church of Martin Luther. Luther urged his followers to resist the temptation to delve into the arrangement of the solar system in opposition to the Bible. Perhaps a similar opposition stopped Aristarchus almost two millennia earlier; that or possibly an air of neglect of his work. In both cases humankind seemed content with the feeling that we are at the very center of things, central to the concerns of God or gods who may have made us in their image, and not out in some astronomical suburb of some sort. A Sun-centered cosmos seemed to shatter our collective ego and self-respect.

The Athenians developed a system of concentric spheres, twenty-six in all, one for each discernible motion of a planet or the Sun or Moon. Eudoxus, Aristotle's contemporary and fellow pupil of Plato, raised the total number to fifty-five, a complex system of interlocking spheres whose gearlike motions drove the only visible parts of it all, the planets themselves. Each sphere was thought to be made of the quintessence, the pure crystalline substance invisible but interlocking with its neighbors in such a way that no vacuum occurred in the plenary universe.

Their Hellenistic successors at Alexandria, perhaps with their new-found Babylonian data, recognized a problem with the Athenian cosmology in that the brightnesses of the planets varied markedly as they danced about the sky. The Aristotelians could not handle this. It had to be assumed that each planet varies in its distance from the Earth and the spheres must be replaced by a yet more complex system of circles or, more correctly, small circles called epicycles wheeling around on larger circles known as deferents. This was the arrangement that was brought to its zenith by Claudius Ptolemaeus, or Ptolemy, in the second century of the Common Era (see Fig. 2.2).

Further, it was known then as now that the Sun takes about a week longer to proceed across the northern half of the ecliptic, from the vernal to the autumnal equinox, than it does to course through the southern half. The exact reason for this was found by Johannes Kepler when he realized that ellipses are the proper orbital shapes; the Sun moves more slowly in summer, for it is proceeding around the portion of the ellipse more distant from the Earth than the other closer half it covers

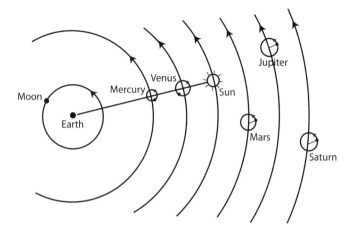

Fig. 2.2 Ptolemy's geocentric system. Each planet moves on a small epicycle, which in turn revolves on the large orbit, the deferent. Note that the system requires the epicycles of Mercury and Venus to be colinear with the Earth-Sun line, and the positions of Mars, Jupiter, and Saturn to be parallel to that same line.

in wintertime. With their unshakable belief that all motion must be comprised of circles because the circle is the perfect figure, all of Kepler's predecessors assumed the Sun to move in a circle, though Ptolemy allowed that the Earth was not quite at the center of the Sun's circular orbit.

Although no one but Aristarchus argued for a heliocentric solar system, his successors made some very significant discoveries. Eratosthenes derived a nearly correct size of the Earth from measures of the altitudes of the Sun and stars as seen from different latitudes. This enabled Greek science to calculate the approximate diameters of the Earth and Moon and the distance between them in their equivalent of miles or kilometers.

Then came Hipparchus, perhaps the greatest astronomical mind of antiquity. Although geocentric in his approach to the heavens, he catalogued the thousand brightest stars and assigned stellar magnitudes to them. We still use his magnitude system today, after increasing its precision.

Hipparchus compared the star positions in his own catalogue with those obtained by an earlier observer and discovered the precession, that great, slow wheeling of the Earth's axis of rotation carrying the stars and sky with it around the perpendicular to the ecliptic once every 25,800 years, a stupendous feat for one of the ancients. Later, about A.D. 150, the magnificent system developed by Ptolemy, geocentric though it may have been, crowned the first triumphant period of science. Each of these steps will be involved in the speculations that follow.

The giant planets, Jupiter, Saturn, Uranus, and Neptune, as well as tiny Pluto, do not exist at all in our proposed triple-star system, and our brightest sun's planetary system ends with Mars. To understand why this is so, we need to briefly describe celestial mechanics and the probable origin of the solar system. We postulate that Osiris lies at an average distance of 23.7 astronomical units, 23.7 times as far from Helios as the Earth, and requires 80 years to complete a single trip around Helios (the average distance from the Earth to the Sun, 93 million miles, is defined as one astronomical unit, or AU). Here we adapt our solar system to the system of the three stars of Alpha Centauri. Thus Osiris revolves in an orbit a little larger in size than that of Uranus. These are exactly the properties of the pair Alpha Centauri A and B. The actual orbit of B about A varies from 11 to 35 AU, a greater swing by far than any planet in our system actually makes. We say that B orbits A, but each star has nearly the same mass; thus in reality A and B revolve about a point a little closer to A, the heavier star, than to B (see Fig. 2.3).

A second mass the size of Osiris would absolutely preclude the formation of any planets around either Helios or Osiris beyond about 2 AU. Its perturbations would be so large that any orbit much beyond that of Mars would be unstable. Any interplanetary material left over after the formation of the stars would be swept up by one of them or ejected from the system into interstellar space to find its way among the other more distant stars. The material that formed Jupiter and the other outer worlds would not have hung around to form these giant planets. Thus the planetary system of Helios extends out only through Mars. And Mars in a system devoid of Jupiter might be a much larger world, more in line with the Earth and Venus. Such a large mass as Jupiter pulls and pulls on everything nearby to the point that nearby rings of debris cannot coalesce into sizeable planets. Still and all, the presence of the much larger mass of Osiris not all that much farther off puts this whole question into an indefinable limbo. Best we leave the terrestrial planets alone here. The implications of a larger Mars will be addressed later on.

Osiris has two small planets of its own, named Zeus and Chronos

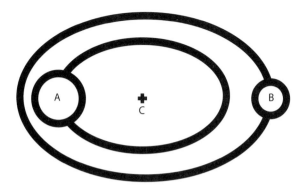

Fig. 2.3 The relative orbits of Helios and the less massive Osiris about C, their common center of gravity. Helios is more massive, thus it moves in a smaller orbit.

after the Greek names for the gods Jupiter and Saturn in the Roman nomenclature. The two lie close to Osiris in stable orbits about that star. Neither is visible to the naked eye, and both were unknown before Galileo observed them with his telescope in 1610.

Osiris is only slightly less weighty than Helios, with around 80 percent of its mass. To satisfy Newton's universal law of gravitation, the two stars must move about a stationary point almost halfway from Helios to Osiris. Each star thus swings about that point known as the center of gravity, or the barycenter, with a period of 80 years, carrying its planets along for the ride. Osiris is about two-thirds the diameter of Helios. Even at its minimum distance from us it subtends a disk of just over one minute of arc. Even then at its closest, Osiris cannot be seen as a disk to the naked eye but appears as a very brilliant point of light. It is 1/2000 the brightness of Helios to us, or about eight magnitudes fainter. Osiris is still nearly six magnitudes or 250 times as bright as the Moon when it is full and at its brightest. Recall that Osiris is a star, not a planet; as such it shines by its own light and is thousands of times brighter than a planet of the same apparent size.

To acquire a feeling for the brightness of the sky when Osiris is up but Helios is not requires a brief diversion. With the Sun, or Helios, well above the horizon, the sky overhead at the zenith is bright, as we are all aware. It is seldom realized just how efficient the eye is at adapting to huge differences in light intensity levels. Hence we are unaware of the great changes in sky luminosity with the setting of the Sun, or Helios. Exactly at sunset, the zenith sky shines at only 1/15 the noontime level. The eye balances this variation, but two photographs of the same view taken near noon and sunset with the same camera settings and exposure time will reveal a vast difference. Civil twilight is the name for the period when the Sun passes from the horizon to a point where it is 6 degrees below the horizon. At this point civil twilight ends, or twilight begins at that point before sunrise. This is about the moment when it becomes time to turn on car headlights and the very brightest stars are just visible. The sky remains a distinct blue if not cloudy, but the zenith brightness is now 300 times fainter than it was at sunset, or nearly 4,500 times fainter than at noon. We can still engage in outdoor activities until about this moment, when the sky is still many times as bright as it would appear on a night with a full Moon.

Since Osiris provides 1/2000 as much light as Helios, the light from Osiris when high in the sky is on a par with that from Helios about 3 to 4 degrees below the horizon. We would think of this as daylight, even when the sky is overcast. Under those conditions too, we are still able to play baseball or tennis or enjoy any other outdoor activity without artificial lighting.

One point arises here in the matter of an underilluminated sky that is sometimes improperly imagined. By underilluminated I mean the sky when dimmer than it is on a normal sunny day. The molecules in the Earth's atmosphere scatter incoming sunlight unequally. This effect, called Rayleigh scattering, favors light of shorter wavelengths strewn all over the sky. The difference between colors that gives each its individuality is its wavelength, blue light being of shorter wavelength than yellow light, and the latter's is shorter than that of red light (see Fig. 2.4). Thus the blue-light component of normal sunshine gets scattered the

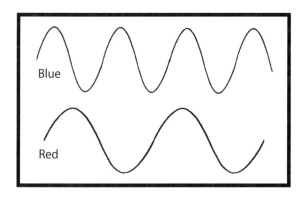

Fig. 2.4 The ratio in wavelengths of blue and red light.

most; hence, the sky appears blue to us. Most of the red light passes straight on through, and the sunlight must pass through a greater mass of air when the Sun is low in the sky near sunrise or sunset. This accounts not only for the blue appearance of the sky, but also the fact that the Sun, the Moon, and the planets and stars all appear redder when low in the sky than when they are seen higher aloft.

It is sometimes incorrectly portrayed in science fiction illustrations that under the fainter illumination of Osiris the sky would appear a deep blue, as it does in some of the paintings of Vincent van Gogh, for example. Consider *The Church at Auvers,* one of the artist's last works (see Fig. 2.5). The sky looks blue, but with an unnatural deep blue color, an effective artistic license but not in the manner in which it is seen at dusk after sunset when still blue, or at noon with a star considerably fainter than the Sun. In reality the color of the sky remains a lighter blue, and the church in this case would not appear a bright mustard yellow tinge but very dark against such a sky, possibly even a nearly black silhouette. We would not experience this combination of dark sky and bright church in the dense air of the Earth's atmosphere at or near sea level. If the amount of air were greatly diminished, if the air pressure were diminished to only a few percent of its sea level pressure, then even at noon the sky would appear as van Gogh has it, an unnatural deep blue with a church still as brightly illuminated as ever. This is the case for the scenery seen in pictures of the lunar landscape taken by the astronauts in the NASA Apollo Program when they were on the Moon. There the sky is jet black since the Moon has no air at all to scatter the sunlight. The lunar surface appears as bright as it does on the Earth because the sunlight there is as intense as it is here.

Mars may pose an intermediate case as its atmosphere is thin, just under 1 percent of the density of the Earth's at sea level. The Martian sky would scatter only a very small amount of incoming sunlight, leaving the sky an inky deep blue color (unless the surface winds on that planet are high and raise dust clouds from the desert areas). Mars is farther from the Sun and receives only about half the sunlight we do, but even if it were at our distance, the effect would be about the same. Those who fly at exceedingly high altitudes above the Earth, far above our commercial jetliners and higher than even the Concorde, may experience the same deep color. The stark brightness of the foreground church in van Gogh's likeness, like objects on the surface of the Moon or Mars, is brought about not by a diminution of sunlight, but by a much thinner atmosphere, unlike conditions at dusk.

Fig. 2.5 Vincent van Gogh, *The Church at Auvers-sur-Oise,* 1890. Paris, Musée d'Orsay, acquired with the help of Paul Gachet and an anonymous Canadian donor, 1951.

The almost full daylight condition would occur whenever the fainter Osiris is well above the horizon unless heavy gray clouds obscure the sky. If we assume Osiris moves along or nearly along the ecliptic that describes the path of Helios among the stars, it must appear alongside the much brighter Helios at one time of year. The much faster moving Helios passes it by and gains on it. Thus about 6 months later it lies opposite Helios in the sky and we would receive near daylight conditions for the entire 24-hour period, much as each pole does under midnight-sun conditions. Osiris moves about 4½ degrees per year along the ecliptic, and thus it takes an extra 4½ days to "catch up" with its changed position. This raises its synodic period to 370 days—in that many days it will again appear exactly opposite Helios in the sky. The entire Earth (except for the polar regions) thus receives about half a day of daylight at one time of year (12 hours at the equator), increasing up to a full day of it over the next 6 months and then back to half in the next half year. The season at which we receive the most hours of daylight would advance by 4½ days each year.

This situation is analogous to the Moon circling the sky each month and repeating its full set of phases in 29½ days. At new Moon the Moon moves right along with the Sun (see Fig. 2.6). We can't see it because its sunlit side is turned away from us and the dark side is lost in the solar glare. By the time it reaches the first quarter phase a week later, it extends the period of sunlight by half the night—some 6 hours; the Sun and the Moon together now shine for 18 hours, three-quarters of a full day. In another week the full Moon rises at sunset and sets at dawn

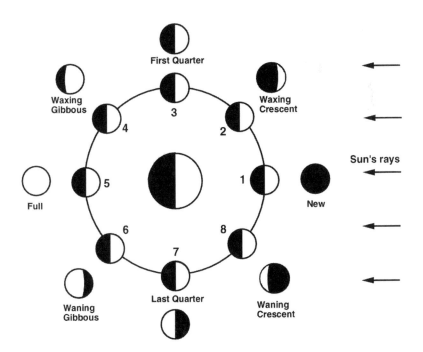

Fig. 2.6 The phases of the Moon. In each position, the Moon appears to us as does the outer ring of circles.

and all hours are bright with sunshine or moonshine. Then at last quarter we are back to an 18-hour bright interval and finally to the minimum of 12 hours at the following new Moon.

Now imagine the Moon to be replaced by a much brighter star shining by its own light. Again we would see daylight increase from about 12 through 18 and up to 24 hours per day, and then shortening back to 12 again. Now instead of a month for the entire cycle we allow a full year plus 4½ days, about 6 months of increasing daylight and 6 more of its reduction. Recalling that Osiris provides illumination about equivalent to that of the Sun just after sunset, we have essentially daylight conditions swelling to a full day—a midnight-sun aspect all over the world, not just near the poles as now.

How would the circadian rhythms experienced by all living things be changed? Would bears hibernate only when the dark period coincides with wintertime? Would nocturnal predators have evolved differently, hunting in subdued daylight as well as by night? What would be our sleeping habits? Would the dark side of human behavior such as nocturnal guerrilla action and lynchings follow a yearly cycle? Our evolutionary and social history might have been very different if the Sun were one member of a double star.

Agriculture and gardening would hardly be affected by the low-intensity light from Osiris unless photosynthesis could occur when it alone is shining, but the times for planting and harvesting might last through much of the night; that is, if we assume night to be the time Helios is below the horizon. The light from Osiris is more than enough to see to harvest crops. Primitive societies would be more assured of harvesting crops in time for the approaching winter. Natural circadian rhythms of people and other animals would need to have evolved entirely differently in many particulars in order to adapt to tolerate twilight levels of light and darkness.

Osiris, when high in the sky, brightens it to an equivalence to the luminosity of the sky when the Sun, or Helios, is rather less than 6 degrees below the horizon, or about 15 to 20 minutes after its setting. This is the "point of equivalence" for astronomers, the moment when the two stars contribute to sky brightness equally. The sky is distinctly blue in color with either star in the sky, and only Venus among the planets and other stars can be easily spotted. Osiris in the heavens means daylight in most cases. In the Koran, day is defined as the time when a black thread can be seen, whereas nighttime begins when only a white thread can be detected. The full Moon may not create daytime conditions in the Islamic world, but Osiris surely would. Stories and fears of the Weald in southern England and trolls in Scandinavia would still exist but would be limited in their temporal application.

Despite the near equality in the masses of the two stars, the ratio between their intrinsic luminosities is greater, being about four to one in favor of Helios. The reason for such a difference derives from the astrophysics of stellar interiors. Early in the twentieth century, when distances first became known for a wide variety of stars, astronomers were able to deduce their intrinsic luminosity in terms of the Sun and each other. We also became aware of the surface temperatures of stars, showing that these temperatures were closely coordinated with the colors of the stars. As one heats metal in a hot flame, one observes that it begins to glow first with a dark cherry red color, but as it heats further the color shifts to orange and still further to yellow and white, and finally, if hot enough, to a bluish tint.

So it is with stars. Proxima is red in appearance because it is cool and barely aglow as stars go, being about 2700° on the Kelvin scale at its surface or about 5000° Fahrenheit. The Sun, or Helios, is yellow-hot at 5800° K whereas Osiris is mildly cooler at 5300° K and orange in appearance. With data like these in hand for many stars, astronomers quickly noted that the orange and red stars come in two distinct sizes. The bright, cool red and orange stars like Arcturus and Aldebaran are truly luminous, but the little ones, such as Proxima, are decidedly smaller, with few red or orange stars between the two groups. The energy output from a star that we see mostly as visible light rises with the fourth power of the surface temperature; this means that a star twice as hot as another (as Sirius or Vega is twice as hot as Arcturus) will put forth a whopping 16 times as much light, heat, and energy per square mile or kilometer of its surface as the cooler star. In order for stars like Arcturus to be so luminous, they must have many more square miles of surface area than Sirius or Vega. Arcturus is known properly as a red giant; it would fill most of the space within the orbit of Mercury. Those stars like Helios and Osiris and above all faint Proxima are also colorfully known as dwarfs. A very few stars are so much brighter even than the giants that they are not improperly labeled supergiants. When the luminosities of giants, supergiants, and dwarfs are plotted against their surface temperatures, they don't fall just anywhere on the resulting diagram. Most, about 90 percent of common stars, fall along a sequence running from the upper left corner to the lower right corner as the diagram is commonly portrayed and appears schematically (see Fig. 2.7).

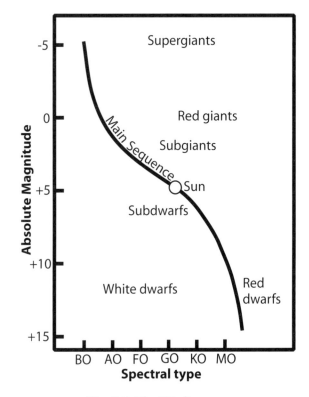

Fig. 2.7 The HR diagram.

This, the main sequence, turns out to be a sequence in mass; the more massive the star, the greater the luminosity by a much larger ratio. In fact the entire diagram forms a study of the evolution of stars. Named the Hertzsprung-Russell or HR diagram after its two discoverers, it is the cornerstone of modern stellar astrophysics and is the most famous diagram in astrophysics. This colorful giant and dwarf terminology has prevailed and remains the same in other languages—*gigante y enano* in Spanish, *Riese und Zwerg* in German.

As discussed above, sometimes we can see the entire Moon when it is in the crescent phase due to the light—earthshine—from a nearly full Earth in the opposite phase from that of the Moon at the time. The Earth as viewed from the Moon is about 40 times as bright as the Moon seen from here in the equivalent phase because the Earth is almost 4 times the diameter and its albedo is several times that of the airless Moon. Even as compared with the Earth at the full phase, Osiris is almost 10 times as bright, and the ratio only increases in Osiris's favor if the Earth is in the gibbous phase, a little way from full. We can be sure that Osiris is fully bright enough at any time to illuminate the entire Moon whenever Helios lights only a crescent segment of the lunar surface.

We have postulated a triple star, one member of which is very faint. In *Nightfall*, one of his most successful stories, Isaac Asimov evokes a sextuple star, six stars all gravitationally coupled with each other. As with any multiple system, stars are found in pairs close together in proportion to any of the other stellar companions. Still, his system moved in such a way that only once in a thousand years did all six stars fall below the horizon as seen from a point on his planet. In *Nightfall*, Asimov's intelligent race had developed celestial mechanics to the point where these rare dark periods were predicted well in advance. Despite this, the inherent panic that swept through their collective nyctophobia, fear of darkness, led them into a frenzied state, during which they tore their civilization to pieces, a kind of terrorism against civilization on a grand scale.

If we were to combine the major premise of the last chapter with this one, we would have three suns, three moons, and six planets, and the solar system would then be without just one of any kind of celestial object. With a selection of suns and moons facing our species, the rise of monotheistic religions might have been delayed by centuries, and even now possibly limited to a small minority of faiths and believers.

The Temple of Venus shines among its neighboring skyscrapers in the New York skyline just across Fifth Avenue from Rockefeller Center. Not the tallest structure in the city but surely the most resplendent, this great silver obelisk, dedicated to the worship of the goddess of love and beauty, stands out among the taller office buildings, gleaming with a lucent shimmer in the morning light of Helios. The mighty shaft is but the most spectacular of the temples in the city; its only nonsecular rival is the pantheon of Mars, a massive, square, red granitic structure situated in lower Manhattan. The great silver shaft of Venus also catches the light of the setting Osiris on its western side, reflecting the softer orange-tinted color of that fainter star, quieter than the yellow-white light from the rising Helios.

Sigmund Freud examines this premise in his study *Moses and Monotheism*, in which he postulates that the great Jewish leader was an Egyptian by birth and heritage. Not long before the time of Moses, in about

1375 B.C., the Pharaoh Amenhotep IV, also known as Ikhnaton, succeeded his father, Amenhotep III, to the throne alongside the Nile River and instituted by coercion a monotheistic religion centered on Ra, or Aton, the solar disk. This Aton religion did not appeal to the Egyptian people, who reverted to their former faiths after his death. The religion of Moses followed Aton closely, and in fact, Freud claims the two may have been identical. Much has been written about this conjecture, and our purpose here is neither to advance it nor refute it but rather to cast the Sun, the source of light and life, in the role of a possible primogenitor of religions limited to a single deity. Glance near the Sun in the middle of a cerulean blue sky (but not directly at it lest your eyes be damaged). What other object in our environment stands out more uniquely than this radiant star? Is it any surprise that this would be taken as a god, if not the only deity, by people just developing agriculture and relying on the Sun and its boundless energy for their continued prosperity?

In his recent bestseller *From Dawn to Decadence,* Jacques Barzun states, "[L]ogically the existence of only one god must mean that all religions are one. Innumerable thinkers, from Voltaire and Victor Hugo to Bernard Shaw and Gandhi, have said so, without much effect on western religious institutions." Our point here is that with more than one sun and one moon, polytheism might have been retained for longer than was the actual case. Perhaps the Roman pantheon of gods survived the fall of that empire on into the Middle Ages and even blended with their Norse equivalents—we can extend the "what ifs" from that surmise into an almost infinite variety of alternative histories. Jupiter or Jove, or Jehovah, might have been first among equals, rather than by himself. Alternatively, Mithras, a Messiah contender complete with his own unique birth, also arose in the first century and might have emerged the winner in the contest between Messiahs in the fading Roman Empire, at whatever epoch monotheism came to the fore. We can only speculate.

3

OUR BACKWARD STELLAR
MAGNITUDE SYSTEM

The stellar magnitude system was developed, as far as anyone knows, by Hipparchus in the second century B.C. He produced a catalogue of the thousand brightest stars visible from his home city of Alexandria. Then he divided the stars into five classes of brightness, in which the first magnitude was reserved for about twenty of the brightest stars. The next brightest fifty or sixty he assigned to the second magnitude and so on through the third, fourth, and fifth magnitude, which comprised the faintest group of stars he could see. Then and now, the number of stars in each successive magnitude interval is about triple the next brighter interval above it. This ratio perpetuates through many fainter magnitudes until we begin to run out of stars in our Milky Way galaxy. The next stellar catalogue of interest and still preserved was compiled by the grandson of the emperor Tamerlane, Ulugh Begh, who worked in Samarkand in the fifteenth century.

Our present magnitude system was first subjected to a well-defined scale about two hundred years ago, but it rests on the system of Hipparchus and is arguably the oldest measurement scheme in use in any field of science. Around 1800, Sir William Herschel, discoverer of Uranus in 1781, his astronomer son John, and a colleague, N. R. Pogson, noticed that a five-magnitude difference was about equal to a 100-to-1 ratio in luminosity, and thus in 1856 did Pogson define the system with precision. We know that the eye and photographic film both record sources of light very nearly logarithmically, not arithmetically. Hence, luminosities in the ratio 1, 2, 4, 8, and 16, for example, will be seen as equally spaced in brightness, whereas the sequence 1, 2, 3, 4, and 5 will not. Luminosities described by the latter set of numbers will appear to be concentrated toward the larger numbers; that is, the light difference from 4 to 5 will appear much smaller than that between 1 and 2. Therefore the magnitude system must consist of ratios, rather than equal steps.

With Herschel's and Pogson's dictum that exactly five magnitudes is set at a ratio of 100 to 1, each step of one whole magnitude will be equivalent to a ratio of 100 to the one-fifth power. Notice that the smaller numbers represent brighter, not fainter, magnitudes. Furthermore, Pogson set the zero point at magnitude 6.0, or 1 percent of the brightness of a star of magnitude 1.0. He chose this limit because it is about the magnitude of the faintest star visible to a keen observer on a dark, clear moonless night. Eyesight varies widely in acuity between people; some very sharp-eyed astronomers can see fainter than the seventh magnitude, and older people lose some of their keenness of vision with advancing age. I am somewhat typical in terms of eyesight. In my thirties I could just see Uranus at magnitude 5.7 on a clear night; now thirty years later I can just spot a nearby double star, 61 Cygni, as one star of magnitude 4.8. My loss of roughly one magnitude over the years is about average.

The stars forming the average of 6.0 were all in the region of the north celestial pole; being visible on any clear night throughout the year, they were accessible to all northern hemisphere observers. The interval between successive magnitudes is thus $(100)^{0.2}$ or very nearly 2.512 to 1.0. A two-magnitude interval marks $(100)^{0.4} = 2.512^2 =$ about 6.31, and a three-magnitude difference becomes $(100)^{0.6} = 2.512^3 = 15.85$. This odd unit, the 0.4 power of 10, is indeed a curious choice. If it were being done over, the choice of a magnitude step would doubtless be either 2 or 10, and the larger numbers would apply to brighter stars. But this is astronomy, where oddball units are commonplace and nice clean ones such as the simple metric units of physics seem not often to be found. That is the curse brought about by huge masses and distances and ancient origins of the oldest science.

The luminance equivalent to magnitude steps is given in Table 1.

When Galileo first employed a telescope for astronomical purposes, he immediately saw stars too faint to be seen by eye, and the natural thing to do was to extend the magnitude range to the seventh, eighth, ninth, and so on. As measurement of starlight improved, subdivisions were put into use; a star halfway between magnitudes 1.0 and 2.0 would be assigned 1.5 and so on into tenths and now even hundredths of whole magnitudes. Similarly, objects too bright to be of the first magnitude were assigned magnitudes of 0, -1, -2, and so on as needed. Thus the brightest star, aside from the Sun, is Sirius at magnitude -1.4. Canopus, the next brightest, shines at -0.7, and eight stars—Vega, Arcturus, Capella, Rigel, Betelgeuse, Procyon, Alpha Centauri, and Acher-

TABLE 1. THE MAGNITUDE DIFFERENCE AS A RATIO IN LIGHT LEVELS

Magnitude Difference	Ratio (to 1.0)
0.0	1.000
0.1	1.096
0.2	1.202
0.5	1.585
1	2.512
2	6.310
3	15.85
4	39.81
5	100.00
6	251.2
7	631.0
8	1,585
9	3,981
10	10,000

nar—all lie between −0.5 and +0.5 and are called zero-magnitude stars (the last two plus the brighter Canopus lie in the deep southern sky and cannot be seen from the latitude of New York). For comparison, most other stars with familiar names, such as Antares, Spica, Regulus, and Deneb, are all first-magnitude objects; Polaris and most stars in the Big Dipper (or the Plough, as it is known in Great Britain) and Cassiopeia are of the second magnitude.

The five easily visible planets all vary in magnitude and are all bright. Venus is of magnitudes −3.3 to −4.7 and, as noted above, outshines every other body in the sky except the Sun and the Moon. It is commonly bright enough to be seen in the daytime and to cast shadows at night. Jupiter is also brighter than any star, varying from about −2.0 to −2.7, and Mars can also be very bright, on a rare occasion reaching −2.8, but fading below +1.5 at most other times. Mercury is usually lost in the solar twilight glare but is very bright, usually between zero and −2, and Saturn covers the range from near zero to about +1.

We have mentioned earlier that Uranus at magnitude 5.7 can just be seen on the clearest of nights by those with sharp eyesight, and Ceres and Vesta, the brightest of the asteroids, can reach the seventh magnitude at times. They and Neptune at 7.8 can all be seen through appropriate binoculars on clear nights. Tiny Pluto struggles along at the fourteenth magnitude and requires a fair-sized telescope to detect. The four big moons of Jupiter discovered by Galileo are of magnitudes 5 and 6 and so would be visible to the eye if the glare from Jupiter, brighter by a thousand times, were not so close to them. The two moons of Mars are at magnitude 11 and 12, but again the glare from nearby Mars makes them difficult to resolve in any telescope. Aside from the Jovian four, the only other satellite in the solar system visible with binoculars is eighth-magnitude Titan, the largest by far of Saturn's crowd and the only satellite to possess a substantial atmosphere (although a number of others are surrounded with thin atmospheres).

The magnitude system, now over 2,100 years old, is admittedly a curious and cumbersome system—it should perhaps long ago have been replaced by another, but astronomers, amateur and professional alike, just cannot break the habit, and magnitudes are likely to hang around for many more years. Their logarithmic nature may not be as ridiculous as it at first appears, for the eye and photographic film both record light sources in approximately a ratio or geometric manner. Try this for yourself—just find a long, straight level street illuminated by evenly spaced streetlights. Notice the brilliance of the first, second, fourth, and eighth lights and you will find that the light ratio between any two of them appears the same, while the ratios between the first, second, third, fourth, and fifth in line appear closer and closer together in luminosity with increasing distance.

Finally, the two brightest of all objects have also been placed in this system. The Moon varies between about -10 in magnitude at the two quarter phases and -12.7 at full, a 10-to-1 ratio. The Sun is nearly constant at around magnitude -27.

We astronomers have added another tier of magnitude usage to our descriptions of the objects in the sky. The absolute magnitude is simply the magnitude a star would have if it were at a distance of 10 parsecs or 32.6 light years. If all stars were lined up at this distance they would differ in luminosity only according to the differences in their intrinsic brightnesses. The magnitude at which we see them, known as the apparent magnitude, incorporates distance as well as luminosity into the bargain. In equations, apparent magnitude is customarily denoted by a lower case m whereas the absolute magnitude is designated by a capital M. This means that m and M for a star differ only by the ratio between its actual distance and the standard distance of 10 parsecs. The equation relating them is $M = m + 5 - 5 \log d$, where d is the distance in parsecs. It is vital to recognize here that if values are known for any two of the variables M, m, and d, the third can be found directly.

4

AN IMPROPER
PROPER MOTION

With the magnitude scale now explained in Chapter 3, we return to the triple-star solar system of Chapter 2.

Proxima, also known as Alpha Centauri C, is one of the tiniest stars in nearby space, truly a 10-watt bulb among stars. It circles the pair composed of Helios and Osiris and their planets at a distance of some 13,000 astronomical units, about one-fifth of a light year. Its orbital period is not well known but must be of the order of one million years assuming, as we do here, a nearly circular orbit. This small star is a very faint red dwarf with a diameter of 0.2 times that of Helios or twice our real Jupiter; its mass is about 0.1 that of Helios. With this size and mass Proxima is far denser than Jupiter and the other major planets. The barycenter between it and our system is about one-twentieth its distance or near 650 AU, 15 times the average distance from here to the real Pluto, our farthest planet. The real Proxima Centauri lies about 2 degrees from the brighter pair in the sky, but being of the eleventh magnitude it still takes a telescope with an aperture of about 4 inches even to see it. In orbit about our system at a distance of some 13,000 AU, it would be visible to the naked eye but just barely. At magnitude 4.5 it would appear as a rather faint star in skies not spoiled by light pollution. It appears as bright as the stars in Ursa Minor (the Little Dipper), apart from Polaris, the north star at the end of the handle, and the two other bright stars at the opposite end in the bowl. Any of the four stars in between these three appears of the same magnitude as would Proxima. It would have been noted and charted by ancient civilizations as just another star, because nothing about it in its appearance or motion could inform them that it is so very close as to be a member of the solar system, gravitationally connected to it.

Since our Proxima is in orbit about us (strictly speaking, about the center of mass between itself, Helios, and Osiris), it does not share the rapid space motion that stars have relative to their neighbors in interstellar space to which they are not gravitationally connected. Unconnected stars move at an average of some 10 to 30 kilometers per second (km/sec) with respect to their stellar neighbors, and share little of the Sun's 20-km/sec motion in nearby space in the general direction of the bright summer stars Vega and Deneb and away from Sirius and Canopus. The Earth orbits Helios at 30 km/sec, but way out there Proxima would creep at a pace of less than 0.3 km/sec in its orbit around the brighter pair of stars, assuming the orbit to be nearly circular. It would dawdle along at a very slow pace for such a nearby star; its apparent angular speed across the sky, called proper motion, is a little over one arc second per year, slower than the apparent proper motions of the bright stars Arcturus, Sirius, and Procyon as well as a few dozen fainter stars hundreds of times as far away. This means that astronomers would have no way of recognizing its connection to our system until the eighteenth or early nineteenth century, when stellar motions of that angular amount could first be detected and measured. Until that time it would have appeared as just another star in one of the constellations along the zodiac if its orbital plane matched the ecliptic rather closely, but for one distinction discussed below.

As early as the fourth century B.C., Aristotle knew enough about the solar system to consider the possibility that the Earth moved. He rejected this hypothesis because he sought to observe the parallax, the back-and-forth motion of the stars that reflects the motion of the Earth around the Sun. He failed to do so and concluded that the Earth did not move. Many others tried to detect the parallax of any star but all failed until 1838, when telescopes were first up to the job. When Nicolaus Copernicus (1473–1543) proposed that we did indeed circle the Sun, he knew the parallactic motion had to exist for his theory to be correct. It was only after that time that the one alternative explanation for failure to detect the parallax came into acceptance. The stars could not be other planets just beyond Saturn; they had to be other stars like the Sun and were so bright and far away that their parallaxes were too small to be detected. Not for three more centuries after Copernicus published his revolutionary theory could this now accepted motion be first measured. But if Proxima was a companion to our double-star system, its parallactic motion would be about 20 times as large as that of the next closest star (the real Proxima, as it happens), and furthermore, unlike any other star, it would move back and forth every six months (as shown in Fig. 4.1) because its motion is almost all parallactic and not

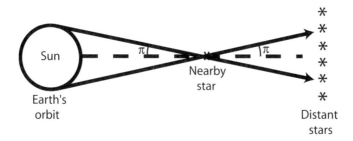

Fig. 4.1 The parallactic motion of a nearby star seen against distant stars. Using the Earth's orbit as a baseline, we see it move by twice the angle pi in a year; the distance goes as 1/pi.

true proper motion at all. It would then have been instantly recognized as being so close to us that astronomers would be aware that it was special, and gravitationally bound to Helios and Osiris and the rest of our system. The million-year oscillation of our system about a point 650 AU off, due to Proxima's gravitational effect, moves so slowly that it would not be noticed until its nature was inferred from the astrophysical properties of our three stars and the close attention the little star would then attract to itself.

How would we have found out just how long Proxima takes to complete one orbit? We haven't been observing it for any million years or even a thousand. Proxima was only seen and charted as a star a few centuries ago, being too far south in the heavens for Hipparchus and other ancients to see and map it. Here, a very trenchant law of planetary motion comes to our aid. Kepler first drew attention to it with his 1619 publication of his third law. It maintains very simply that the square of the period of a planet is equal to the cube of its mean distance from the Sun, nothing more, thus $P^2 = a^3$ where P is the period in years and a is the average distance in astronomical units. For example, a planet with a period of 8 years will be found in an orbit of mean distance 4 AU, because $8 \times 8 = 64 = 4 \times 4 \times 4$. This doesn't quite tell the whole story; it was Newton who realized the force behind this law, which requires that the masses of the two objects in mutual orbit be considered. Newton recast Kepler's equation to read $P^2 = a^3/[M \text{ (Sun)} + M \text{ (planet)}]$; that is, the sum of the masses must be taken into account. The mass of the Sun is so much larger than even Jupiter that Kepler had no means to detect the differences. Newton derived the complete expression from his work on gravity and also demonstrated that it was universal; it applied everywhere. Only then was the influence of the differing planetary masses first noted.

The Sun's mass is one solar mass by definition, and Jupiter's mass is just over one-thousandth of this, hence the correct term for the two in Kepler's original version is swelled by a factor of 1.001, hardly detectable in Kepler's day. But in our description of the Sun as the brightest of three stars, the stellar masses impose much larger changes. The Sun or Helios has a mass of 1.0 but Osiris adds another 0.8 solar mass; thus their mean separation, with P found from observation to be 80 years, is $a^3 = 80^2 \times 1.8 = 22.6$ astronomical units, close enough to our result in the last chapter of 23.7, given that our figures for the two masses are approximate. Proxima's period around the brighter pair is just the square root of $13,000^3/1.9$ (we must use 1.9, not 1.8 for the sum of the masses because now Proxima's small mass of 0.1 must be added to the total). The period of the little star works out to be 1,075,000 years, for which an even one million years is a satisfactory representation, given the observational uncertainties involved.

Now with the conjecture that our Proxima is so close as to orbit the solar system, we surmise that back in 1826, when in our version the German mathematician-astronomer Friedrich Wilhelm Bessel determined the true distance of this small, red dwarf star from its parallactic motion (as he actually did for a more distant star 12 years later), he had first to realize that its seeming proper motion oscillated with a period of one year, rather than moving in a straight line. In the dozen years until the parallax of any other star was measured in our real sky, Bessel came to understand that its true proper motion, the true angular motion across the sky, was much smaller than the parallax. For no other star could the parallax be so much larger than the proper motion unless it was moving directly toward or away from us. In any large sample of thousands of stars, perhaps one or two might meet these qualifications by chance, but the infrequency is sufficient to assure us that randomness is not the cause here. In the case of Proxima, however, the parallax is more than 20 times as large as that of any other star, and the chance that it was an unassociated star with its primary motion component directed right toward or away from the solar system would be embarrassingly small. Bessel correctly recognized the other reason for this discrepancy, that it is in orbit about us and therefore shares the major component of our own motion among the stars. The motion of our outer planets is much the same in kind. Pluto, for example, requires about 247 years to circle the Sun just once. It therefore has an apparent proper motion of about $1\frac{1}{2}$ degrees per year. But it is on average about 40 astronomical units from us and the Sun, and our orbital motion is superimposed on the linear motion and shifts Pluto by nearly 3 degrees back and forth in an eastward and then a westward movement we call retrograde motion. Hence, the parallactic component of its apparent motion is about twice its proper motion across the heavens. All of the other planets reveal this cyclic component as well; Mercury and Venus do so when they swing westward back across the Sun as each in its turn overtakes the slower Earth on its way by.

Proxima moves in just this same manner and could be taken for a very, very distant planet. If it were not gravitationally bound to the rest of us but just "passing through," it would hurtle by us at typically 20 or 30 km/sec. The star would appear to possess a proper motion of well over an arc minute annually, shifting by half a degree (the Moon and Sun's apparent diameter) in just about 20 years—any star chart maker would be aware of such a speed in no time. This would have been noticed by some of the ancients, and Aristarchus might well have the proof he needed to point out that stars are not all plastered on some celestial sphere but are instead hanging out there at different distances, and, of greater importance, support for his contention that the Earth orbits the Sun. Pluto under the same circumstances would sail along at several astronomical units per year and fly out of the system in a matter of decades, unnoticed unless and until modern large telescopes became available.

As it is, there is no motion of stars in the sky, other than their mutual reflection of the Earth's rotation that can be seen over a short time interval. Neither the precession of the equinoxes nor the proper motion of any star can be detected within a single lifetime; these deliberate motions can be identified only over centuries, which means that record-keeping and some degree of literacy must be present. I may have reduced the detection of proper motion to as short a period as possible with low magnification when I observed the bright star Arcturus 50 years ago and again this year with a small telescope of about 30 power. I could detect its change in position relative to other stars nearby over that interval, but only with the foreknowledge that it is a rapidly moving star. Never could I have noticed the shift either with the naked eye alone or without the prior knowledge of its high velocity. The apparent rapid motion that Proxima would exhibit would be unique in this respect.

Twelve years later, in 1838, Bessel was one of three astronomers who independently measured the parallax of another star not connected to our solar system. He chose a faint star in the constellation of Cygnus that he knew had a large proper motion and therefore was likely to be very close to us as stars go. At the same time Wilhelm Struve measured the parallax of Vega, and Thomas Henderson in South Africa determined that of Alpha Centauri, still the closest star system at 4 light years, with the largest parallax of all. This constituted the final proof of the heliocentric theory of Copernicus, as modified by Kepler and Newton, nearly three centuries after its widespread acceptance. Even now in our own times, the present pope, John Paul II, has reiterated the acceptance of it and Galileo's support of it, reminding us again that whenever religion or government attempts to dictate or legislate science, there is bound to be trouble, a reminder that seems to need repeating over and over again.

5

ALL OUR YESTERDAYS

Tomorrow, and tomorrow, and tomorrow,
Creeps in this petty pace from day to day,
To the last syllable of recorded time;
And all our yesterdays have lighted fools
The way to dusty death.

WILLIAM SHAKESPEARE

The tower stands foursquare to the ocean and the wind. It is the pride of the city and the Empire. The city, Alexandria, the second largest in the Roman Empire after Rome itself, stretches for miles along the juncture between one of the mouths of the Nile River delta and the seacoast. It is an important granary and supply depot for lands surrounding the entire Mediterranean Sea. With its renowned library, housing scrolls with all of the knowledge of the world, the city is also its intellectual center.

A young man gazes out over the harbor at the tower with a mixture of pride and wonder; pride that such a marvel could be built, and wonder that it stood at all. At over 400 feet, this Pharos, this giant lighthouse, is the tallest structure in the world along with the great pyramids, not far to the south. For another millennium and more, it would stand until humbled by an earthquake.

His name was Claudius Ptolemaeus and he bore the illustrious name of the royal family of pharaohs that had ruled Egypt under the name of Ptolemy until the death of Cleopatra some 130 years earlier. Now, at eighteen, with a name that no longer connoted royalty, he gloried in his city, its library, and its remarkable lighthouse.

It seemed to him that he could sit on the pier and watch the city and harbor forever, never tiring of it. The ships came and went incessantly; the unloading of their cargoes was carried on with dispatch along the piers that stretched the entire length of the city, fanned by the eternal mix of its residents from the dark-skinned Nubians of the headlands of the Nile River far to the south, to the fair Bretons from the Hibernian Islands lying in the great ocean off the northwest coast of Gaul.

Claudius searched for the evening star and spotted it in the clear blue afternoon sky—he was well aware that this star, brighter than any other, is visible before sunset. It is the famed goddess of love and beauty known as Aphrodite, or Venus, as the Romans addressed her. Later, after sunset, she would grow ever more brilliant in the waning dusk until she cast shadows, just like the Sun and the Moon. As darkness fell he could see the evasive Mercury below her and the Moon above. For one brief moment he would realize these worlds and ours as the renowned astronomer, Aristarchus, had pondered them, as circling the brilliant Sun, with our planet as one of them in their midst with an orbit between Venus and Mars . . . no, he shook this tantalizing image from his mind. His other illustrious predecessor, the great astronomer Hipparchus, had considered and discarded this radical step—as Ptolemy will also. Over the next half century this young man would become the greatest scientist of antiquity, forming the most original and true picture of the great cosmos, the astronomy and astrology of it, while mapping our Earth along the way.

In the preceding chapters, we have postulated modest alterations in the design of the solar system that might have led to the introduction of a heliocentric system early in the development of the Roman Empire, affecting it and the post-Roman European civilization. If we could have detected Jupiter's moons without a telescope as the sharp-eyed among us almost can; if Venus subtended a larger disk with phases discernible to the unaided eye; if the writings of Aristarchus had been accepted more thoroughly by Hipparchus than we think they were; if the Moon, Mars, or Venus had a visible moon in orbit about it; a Sun-centered cosmos might have been an integral part of the creed of the Western world and the religions within it. In this chapter we postulate just two differences between the world that is and the world of the might have been. First, Hipparchus accepts Aristarchus's major premise (of a heliocentric system), and second, the telescope in a simple form was developed at the height of the Roman Empire, perhaps in Alexandria. Both are just plausible within the scope of our present knowledge.

We are sadly ignorant of the contents of the great library at Alexandria and can only speculate on the memoirs of the early Hellenistic scientists it might have contained. It is also uncertain when and where the first lenses were ground. Certainly glass was available in the classical world, but references to optical aids of any kind are spotty and uncertain. Some of the ancients, from Aristophanes to Pliny and possibly Ptolemy himself, made reference to transparent round objects that may or may not have been sufficiently lenticular in form for two of them with differing focal lengths to be aligned such that they would form a simple telescope and magnify a distant object. Here we assume that they did so, and with that in place we proceed.

The lifetime of Claudius Ptolemaeus, known to posterity simply as Ptolemy, coincided with the zenith of the Roman Empire. His accomplishments transpired under the reigns of Trajan, Hadrian, Antoninus Pius, and perhaps also Marcus Aurelius. He lived mostly or entirely in Alexandria at the mouth of the Nile River in Egypt.

Ptolemy's dates are not known exactly, nor are the factors that motivated this genius and preeminent scientist of his time. He is presumed to have been born about A.D. 90 and to have flourished during the middle of the second century, making observations that can be dated to the years from around 130 until 150 or later. His great treatise, the *Syntaxis*, known more widely by its Arabic name, the *Almagest*, a word meaning "The Greatest," must have been written around 150. In it he laid down the most sophisticated and mathematical world model before Copernicus.

According to Owen Gingerich in *The Eye of Heaven*, "It is difficult to convey the elegance of Ptolemy's achievements to anyone who has not examined its details. Basically, for the first time in history (so far as we know) an astronomer has shown how to convert specific numerical data into the parameters of planetary models, and from the models has constructed a homogeneous set of tables—tables that employ some admirably clever mathematical simplifications, and from which solar, lunar, and planetary positions can be calculated as a function of any given time. Altogether it is a remarkable accomplishment, combining in a brilliant synthesis a treatise on theoretical astronomy with a practical handbook for the computation of ephemeredes."

Little else is known of the life of this most fecund of ancient scientists, his accomplishments and failures—far less than of many of his contemporaries in other disciplines. Vastly more is known of the lives not only of the Caesars, with their historians Suetonius and Tacitus, but also other prominent ancients from Plato to Augustine of Hippo, from Pericles to Constantine, than is known of Ptolemy. For such a commanding period of history, it is perplexing that he remains so much a shadowy figure. Only his literary works have come down to us; in addition to the *Almagest* he wrote the *Tetrabiblios,* a treatise that set down the principles of Western astrology that still dominate much of the current practice in the field, as well as another work on the geography of the known world. Had he recentered the universe at or near the Sun, might he have been better acquainted with one or more of those illustrious emperors at the epicenter of power and influence in the Roman Empire, and might his personal memoirs have survived in some form? Might astrology have been eclipsed by the foreknowledge that the Earth is just another planet circling the Sun between Venus and Mars and that the stars are other suns at distances incompatible with our smaller solar system?

The most perplexing problem facing astronomers from Hipparchus to Ptolemy, a span of nearly three centuries, was the motion of Mars. The red planet defied every attempt to account for its retrograde motion, a trait shared by all planets farther from the Sun than the Earth, the so-called superior planets. When nearly opposite the Sun in the sky, each of these superior bodies stops its normal eastward motion along the ecliptic and through the zodiac and veers toward the westward or retrograde direction, counter to the dominant prograde direction. After a few months it again stops, only to resume the normal prograde motion. Not only does Mars make a larger retrograde loop in the heavens than the more distant planets, it does so unevenly; the loops are twice as large in the portion of the zodiac near Cancer as they are in the opposite direction near Capricorn. The explanations of these well-observed phenomena defied every attempt to account for them, albeit with lessening error from classical times through the Arabian supremacy in astronomy, up to Copernicus and even Galileo. It was Johannes Kepler who finally explained them correctly in his three laws of planetary motion, published in 1609 and 1619, when he broke the paradigmatic mold by rejecting circular orbits altogether in favor of elliptical ones. He had Tycho Brahe's observations to work with, the best pretelescopic positions ever observed and recorded. But of greater moment was Kepler's realization that orbital motion among the planets was not governed by their positions or geometry but by a force that appeared to be centered at the Sun. Kepler's was an entirely new concept of the mechanism of the system that Newton expanded nearly a century later, using his great knowledge of the underlying physics.

Had Ptolemy accepted and demonstrated a cosmos centered at or near the Sun, he would have been a more central figure. What might have prompted him to take this bold step? Here it is assumed that he was influenced in this, as in so much else, by Hipparchus, his very worthy predecessor, and by the availability of the new gadget composed of just two lenses of different focal lengths and a tube in which to support them. That would place Ptolemy in the same position as Galileo in 1610; Ptolemy would have been the first to make that string of discoveries that revolutionized human thinking about its environment. The phases of Venus and the moons of Jupiter doubly confirmed that the Earth is not the center of all motion. Ptolemy already believed that the Earth was not at the center of the circular orbits of the other planets; his off-center circles are almost indistinguishable from nearly round ellipses, but the paradigm change required for this bold step may have been beyond anyone of the time.

Still, a heliocentric system of Ptolemy would have been a near copy of the one Copernicus developed fourteen centuries later. Might this have spurred Hellenistic or Roman astronomers to anticipate Kepler and even Newton at the time? Even a partly positive answer to this would have changed our world in many unforeseen ways. The Greeks and Romans did not develop a strong tradition of quality observations of the motions of the planets, as did the Babylonians and their Muslim successors. The blend of the observations of these groups with Greek science fueled the later Scientific Revolution. Ptolemy had access to some of this material; with a Sun-centered model, he or a successor just might have stumbled upon elliptical orbits. Circular motion as the perfect shape and thus the only acceptable figure for an orbit was not as securely entrenched then as it became later with the rise of the hegemony of the Christian Church and its strict adherence to scripture. A window of opportunity opening during and shortly after Ptolemy's lifetime just might have allowed the discovery of Kepler's laws, although the mathematics for his level of sophistication were not yet fully known.

The sense of place, so dominant during the Middle Ages, would surely have been shaken, if not convulsively undermined. The elegant system of Ptolemy might not have lapsed back into the simpler metaphysical one of Aristotle, allowing dogma to prevail over rational thought, with mathematics taking a back seat for centuries until its reemergence through an infusion from the Middle East. Perhaps Bede, even living in boreal obscurity as he did, might have persuaded Europe to remodel its calendar to include the temporal dislocation he knew to exist in the Julian scheme in his own time, and not eight centuries hence with Pope Gregory. Might the highly mathematical Gerbert at the end of the first millennium have ended with a more quantitative cosmos along with or instead of the papacy as Sylvester II? With no convenient locations for Heaven and Hell, might Dante have remained a more obscure poet? Much might depend upon just when the West transformed from a life of the spirit and the afterworld into one emphasizing the here and now and this material world, as Barbara Tuchman states so eloquently in *A Distant Mirror*, her depressing account of the fourteenth century. With a heliocentric system in place, the activity during the thirteenth century of thinkers like Robert Grosseteste (ca. 1168–1253) and Roger Bacon (ca. 1220–1292) might well have resolved the seeming paradox whereby on a rotating Earth an object thrown aloft might fall well to the west of the place where it was thrown. They might have made the strides necessary to command the concepts of force and momentum to this point and their correct response. That century brought the rise of the first universities and the translation of Aristotle and other ancients from Greek and Arabic into Latin, the lingua franca of the age. Might the Renaissance have been born then in Paris and London, two hundred years before it arrived in Italy?

These speculations, fanciful or not, appear to face their biggest hurdle (in this seeming rush to bring on our modern world centuries ahead of time) with the genius of Isaac Newton. For its full emergence, his triumphant new world system seemed to demand the insight only he could provide. Not for nothing do he and Einstein stand alone in science. Without the one or the other we would, in time, have come upon the physics we now know, but it might have been with a great diffusion through a number of empiricists plodding their way to the modern universe through experiment, and not through grand conception later verified by observation.

And if the Scientific Revolution and the post-Newtonian world had come on a century or two earlier, it is possible to consider an earlier Industrial Revolution. A premature comprehension of the electromagnetic spectrum and its governing rules, as well as gas laws, might have induced everything from the steam engine to the dynamo at a much earlier epoch. The institution of slavery required the presence of machinery and its attendant economics, as well as its long-standing moral reprehensibility, to bring it down, but the two together finished it, at least as an institutionally approved practice. What if that had happened generations earlier—might the American Civil War not have occurred? Might our great urban agglomerations with their populations in the millions have begun, not with London in Victorian times, but much earlier, perhaps still with London in the lead? Jane Austen, rather than Charles Dickens and George Eliot decades later, might have been among the first to introduce industry and the railway into her novels. The works of Mozart and Beethoven might have been recorded in their own era at the time of their composition. Faraday might have displaced Maxwell in leading us to the comprehension of electromagnetic radiation that formed the impetus for Einstein's special theory of relativity.

Currently it has become fashionable to consider events in a multicultural milieu. What would be the effect of an acceleration of the Scientific Revolution on non-Western cultures—might it have placed them at a further disadvantage, or could they have benefited more from the great unique wave of Aristotelian metaphysics and other Greek science far earlier than in fact they did? If China were to have come by the Greek tradition sooner than it did, it might well have mechanized the great Chinese armadas of the early 1400s and truly conquered the known world—the possible alternative consequences of such a move are endless.

6

WE ARE ALONE

The Sun was setting on that bleak November day. In another hour the sky would be dark. On overcast nights we could see a modicum of light from natural and artificial illumination from the ground, but on clear nights the world and the sky are totally dark. No moons or planets act to illuminate the night because none exist. Only the stars remain. On occasion the Aurora Borealis and Aurora Australis shine near the magnetic north and south poles, respectively, in our upper atmosphere. Our spacecraft venture only into inner-earth orbits since there is no need to go farther into the black void. Physics is a rich field but astronomy is not; even the word, along with its erstwhile soulmate, astrology, appeared rarely. Both words are based on astri, *the Latin word for stars, but in this world there are no measures for the stars.*

Here we assume a solar system devoid of everything except the Earth and the Sun. No other planets move about in the sky and no moon accompanies our world because no Mars-sized blob of condensed matter crashed into the proto-Earth to wrench from it the material that later consolidated into our Moon. How likely is such an arrangement? I leave to those who work with models of newly formed planetary systems the question of whether a gas cloud can condense into a single star and a single planet with no significant residue left over. Now that dozens of large Jupiter-sized planets have been clamoring for our attention by tugging on the primary stars they are orbiting and pulling at them like a small dog on a leash, we find that most stars seem to be members of binaries or multiple star systems; it may be that such a lone planet circling a lone star may apply in fact to only a tiny minority of cases, if any.

The sky would appear no different from the sky we see on many a clear night whenever the Moon is not above the horizon. Every night would appear the same except for the slow seasonal changes due to our revolution around the Sun; thus, the winter constellations would in time give way to those of spring and later the ones of summer and autumn. No visible planets would amble across the sky, but none are visible on many nights now as it is. But despite this mutual common-place, we will find that the bunch of other planets formed along with the Earth and the Sun 4.6 billion years ago have enriched our lives in many ways.

Perhaps the most significant difference occurs in a consideration of the history of science and ideas in general. For example, astrology might have been discarded centuries ago because it would at most be limited to the simplistic sun-sign variety, the kind that appears in most newspaper columns. No Saturn would be lurking in the seventh house to complicate the simple horoscope; no Jupiter would be in quadrature with Mercury, ever ready to forecast momentous events, good or evil. Astrology at this sparing level might never have attained its present high degree of popularity. However, charlatans would still be able to manipulate many other pseudosciences ready to fill its place; tea leaves, phrenology, and the tarot cards would still be available to attract the gullible and their money.

The science in astronomy also takes a completely different turn. From classic times, when Pythagoras explained the nature of eclipses and shortly thereafter, Aristotle and Eudoxus imagined a solar system of interconnected shells accounting for each planet's motions, the planets and the way they are seen to move became the principal medium in which modern science was born. Neither their geocentricity nor later the more complex cosmos of Ptolemy, nor the heliocentric planetary system introduced by Nicolaus Copernicus could be imagined. Whether the Earth or the Sun did the moving would remain a moot point until very late in the post-medieval world. Only when the aberration of starlight and the direct measure of stellar parallax could be mastered using the improved instrumentation in the eighteenth and nineteenth centuries could the problem be settled. The astronomical unit is by far the most important distance in the universe, since the measurements of distances to every object within or without the solar system, save only the Moon, all depend on its accuracy. Yet it could hardly be well determined even today, since in our present world we measure the distance not directly to the impossibly brilliant Sun, but to other planets and nearby asteroids, ever since Kepler established his three laws of planetary motion, giving us a relative scale for the distance to the Sun and each planet in the solar system. Even knowledge of the speed of light might have been delayed, as it could only have been determined in the laboratory.

Could Isaac Newton have established his laws of motion and gravitation in a system constrained to just two bodies? He would have no moon to compare with a falling apple, and no way to derive the distance and diameter of the Sun alone. Nor would Newton have had any way to derive the mass of the Sun, its most important physical parameter. In the late nineteenth century the Earth's mass would be measured in the laboratory, but that of the Sun, some third of a million times ours, would be unknowable until G, the gravitational constant, was evaluated, and this would not be likely until space vehicles were placed in orbit about the Earth and the Sun.

The diameter of the Earth would have been well known—the Sun and stars are sufficient for that and for celestial navigation. But only the recognition that the Sun is a star would spark a crude idea of its size in terms of stars in general, and then only after a dose of comprehension of the internal constitutions of the stars. Later, without knowledge of a single stellar mass, the astrophysics involved in stellar constitution and evolution could not be pieced together as it has been over most of the last century. Newton might not have known the proper shape of the Earth's orbit. With no reference points and no 30 years of planetary observations from Tycho Brahe, how would Kepler have shown that ellipses—not circles—fit the data remarkably closely? Astronomy would have lagged behind the physics of Galileo and others, but even his accomplishments might have been limited to studies of the pendulum and other laboratory equipment with no moon or planets, since his telescope could not magnify the stars into disks. Finally, no comet would have been available for Edmond Halley to prove that comets orbit the Sun as planets do and to demonstrate the universality of Newton's laws. With a shorter day and a slower rate of precession, the polar wobble shifting the positions of all of the stars (since tidal force is limited only to the Sun), the first detection of the motions of individual stars might still have been Halley's recognition of the proper motions of Sirius, Procyon, and Arcturus from a comparison of his positions of these bright stars with those Hipparchus described almost 2,000 years earlier. Only then might we suspect that the stars were not all at the same distance, as if pasted onto a transparent spherical shell.

Newton realized that his law of gravitation required the premise that the gravitational field of an object would be unchanged were all its mass concentrated into a single point at its center of gravity. Thus the attraction of the Earth on us is identical to that from a single point of the Earth's mass located at its center, almost 4,000 miles beneath our feet. To prove this concept, Newton needed calculus, so he created and developed it! Had he not done so, his contemporary, a German mathematician by the name of Gottfried von Leibniz, would have, and in fact did, come up with it independently a few years later. Even so, had Newton not had the impetus to do so, calculus might not have been applied to gravity and orbit theory for any number of years—no orbit but this indeterminable one of our own world about the Sun would have been known at least until William Herschel's work on the orbits of double stars a century later.

No week or month would suggest itself to us; we would need to turn perhaps to the menstrual cycle or gestation periods for units and divisions of time between the day and the year, or do without. No eclipses occur, either of the Sun or of the Moon since there is no Moon. Eclipses, particularly total solar eclipses, are so rare at a given location that their encounter in historical documents would be a bonanza for the historian and the archaeologist. We now know the exact dates of all eclipses that have occurred in recorded history, and their use in the calibration of radiocarbon dating and other techniques of age evaluation has been helpful in securing a correct chronology for civilization's formative years. Without eclipses, the problem would have been more difficult, although with the availability of tree rings, maybe not impossible.

With no other planets about, the Earth might not be much different than it is now. But if the Moon weren't there in its customary place, life would indeed be far different. Without the lunar tidal drag on our oceans to slow its spin, the Earth's motion would have remained far closer to its original high rate of rotation, revolving every 8 hours or so. The Sun has a tidal effect, but the proximity of the Moon makes it the much stronger influence. With such a high rate of rotational velocity, the Earth becomes a much windier world. The whole climatic scheme is driven by the rotational speed and the prevailing westerly winds caused by the temperature gradient lying between the tropics and the polar regions. The balance between the two effects gives rise to the jet streams that channel the weather near the surface. Evolution might well have developed higher orders of life in a breezier place, perhaps even intelligent life, but they would not be as they are in this world.

With no moon, the long-term climate might have been vastly

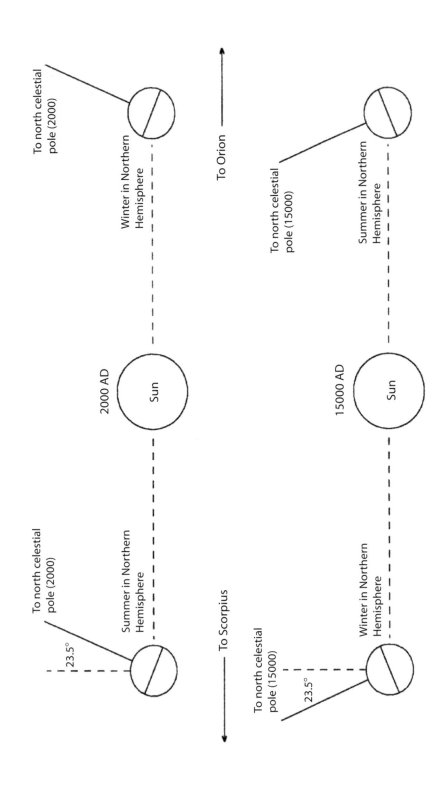

different. Three long-period variations in the orientation of the Earth and its orbit account for most of the ice ages, the glacial periods that have dominated the Pleistocene Epoch. These are known as the Milankovic variations, named after the Serbian astronomer who discovered their climatic influence. One of these is the precession wherein the gravitational influence of the Sun and particularly the Moon cause the Earth's axis of rotation to gyrate slowly, like a gyroscope or a simple top. As the latter slows down, its axis tilts from the vertical and precesses in a slow motion to the side. The familiar 23½° inclination of our axis of rotation to the plane of our orbit, known as the plane of the ecliptic, causes this sideways spin, whose period is 25,800 years (see Fig. 6.1). In that length of time from now the north pole will again point toward Polaris, the north star, after passing near several other stars, each of which will serve its turn as the pole star.

The second slow change is in the angle of the tilt itself. Over a period of 41,000 years, the inclination varies from about 22° to more than 24½° and back again. We know this in part from historical times; Stonehenge, erected 4,000 years ago, was aligned for an inclination of 24° (its proper value at that time), and since then the angle of tilt has lowered by about half a degree. Finally, the eccentricity of the earth's elliptical orbit varies from less than 1 percent to nearly 5 percent (presently about 1.7 percent, or pretty nearly round) in a random and possibly chaotic manner over hundreds of thousands of years. The Moon figures in all of these slow changes; without it our precessional period would be closer to that of Mars—177,000 years. Given Mars' greater distance from the Sun and the absence of sizeable moons, this much longer period of precession comes as no surprise. With no moon, our precession might be of the order of 100,000 years.

The variation in the tilt of the Martian axis is much more alarming. This angle wobbles erratically from about 13° to over 35°, with a present value that happens to be close to that of the Earth. Such a sway would mark our world too, exaggerating the seasonal variance and wreaking havoc on any life systems that might evolve on this planet, as on Mars, if our Moon were not there to stabilize this variation. The orbital eccentricity of Mars is at present about 9 percent, almost twice the maximum eccentricity of our orbit. Life and perhaps intelligent life could form on this windy, wobbly moonless version of our familiar

Fig. 6.1 (*opposite*) The precession shifts the alignment from the upper case of A.D. 2000 to the lower case of A.D. 15,000.

Earth, but its chances would be slimmer. Perhaps this is why life on Mars, if it exists at all, seems surely limited to simple creatures of the single-cell variety.

Albert Einstein is one whose career would have been far different in our abbreviated model of the planetary system from the illustrious one we know. Einstein conceived and published his theory of general relativity in 1915, and although he was not particularly motivated by experimental confirmation of it, he suggested three observational tests by which it could be validated. The first subjected to observational examination his concept that light rays are bent by gravity. The Sun, alone in the solar system, possesses a gravitational field of sufficient strength for this purpose. In 1919 a total solar eclipse provided the opportunity for the test, since stars could be observed near the solar disk only on the occasion of a total solar eclipse. Photographs were taken during the eclipse and again some six months later, at night when the Sun was on the other side of the sky, and the positions of the stars were measured and found to have been displaced *away* from the solar disk at the time of the eclipse when compared to the empty sunless star field.

Sir Arthur Stanley Eddington led an expedition to the eclipse site, and after measurement of the stellar images photographed then and later, his team found a discrepancy of about the amount predicted by Einstein, despite some errors in the observing process. When the findings were released, the deviation of starlight by the gravitation of the Sun was confirmed and Einstein at once became a world celebrity.

The second and only other test of the theory that could be adequately confirmed at the time involved Mercury and its orbit. Under the prevailing system, planetary orbits themselves rotate about the Sun due to gravitational perturbations from the other planets. Mercury is close to the Sun and moves at a great speed in its orbit about it. Furthermore the orbit is of a comparatively high eccentricity, and the movement of the orbit is easy to detect. While all other planets reveal this property, none do so to the readily observable degree as does Mercury. Imagine bending a coat hanger into an ellipse representing Mercury's orbit and placing a bead on the wire to represent the planet. The bead would whip around at a high velocity as Mercury does, ineluctably circling the Sun every 88 days. But Newtonian mechanics predicts that the orbit will also spin around, although at a much slower rate. With the great increase in precision brought about by the telescopes developed in the nineteenth century, the rotation of Mercury's orbit could be measured and was found to amount to 574 seconds of arc per century—in 2250 centuries the orbit would return to its original position, having traced out a rosette in the process. After all of the gravitational perturbations of the other planets were accounted for, the Newtonian model predicted only 531 arc seconds per century. The discrepant shortfall of 43 arc seconds was not accounted for under the existing paradigm, but general relativity predicted an extra amount of just this size. Einstein was once again right!

In our postulated model, however, there is no Mercury orbiting the Sun and no moon to eclipse it as it does, allowing stars to be visible near the solar disk. Neither of the two tests proposed by Einstein would have been possible or even imaginable in our stripped-down version of the solar system, and Einstein would surely not have been the celebrity he became and would not have been named the Person of the Century in 2001 by *Time* magazine—that honor would settle on one of the two runners-up, Mahatma Gandhi and Franklin Delano Roosevelt. Or might it have devolved upon Adolf Hitler as the most influential person of the past century for good or evil? In the years since, Einstein's theory would have been confirmed, but likely not during his lifetime.

But, above every other condition of a system constrained to just two objects, we sense a loneliness, a haunting void, like an empty house, furnished but without the comforting presence of a family to provide the natural companionship we humans so keenly feel and cherish. The first space probe that leaves the Earth beyond a near-Earth orbit must roam through space for 4 light years or 270,000 times as far as the Sun to rendezvous with another star, hopefully one with planets to land upon. It promises to be a desolate, centuries-long journey, without much hope if planets are not discovered there beforehand. Would we feel this remoteness if we knew nothing else? Would we feel deprived of other life with which to share this vast universe? Our neighboring planets may or may not harbor life, but we like to think that between them they have or at least could have produced life, and even consciousness. Their very presence portends other richly endowed planetary systems; without them we are likely to feel very much alone.

At the collapse of the Roman Empire came the transformation of Mediterranean Europe into the early Middle Ages. These so-called Dark Ages were soon accompanied by a cleavage of the lands surrounding the Mediterranean Sea into a southern shore that quickly became Islamic, and, on the north side, a Christian Europe that turned its center of activity away from Italy and the south to the Frankish-Germanic lands of the north. The reigning scheme of Ptolemy gradually gave way in the West to the simpler, less quantitative one proposed by the Athenians. Many factors played a part in this; the Church was more favorable to simpler models in order to emphasize creation according to Genesis and to promote the central location of mankind as the summit of His creation by dint of our home at the exact center of it all, not slightly removed from it as Ptolemy had postulated. Learning in Europe during this period was limited mainly to monasteries, in which it was broken down into seven topics.

The first three, dubbed the trivium, dealt qualitatively with discourse and the three disciplines that thrived on it: rhetoric, grammar, and logic (dialectic). The quadrivium contained the four others, in which quantitative thinking has a place. These are astronomy, arithmetic, music, and geometry. All four were based on scientific principles even then, and inevitably, in the breakdown of learning centers with the disappearance of classical civilization, these were the subjects that suffered. Teachers who understood and could teach them gradually disappeared; for this and other reasons the students left behind, with few exceptions, learned only of the trivium.

In many current textbooks for introductory courses in astronomy, the 1400 years between Ptolemy and Copernicus, if discussed at all, appear as a gap in which nothing happened and astronomy was on hold throughout this entire period. Nothing could be further from the truth. Western society paled in these Dark Ages while the recently developed Islamic world picked up where the Greeks left off, with one major difference. They furthered their own deserved reputation for collecting observations of quality far beyond anything in the Greco-Roman world. But the thread, the meaning, of science went unimproved. The Middle East, far beyond the Europe of the time, produced commentators who preserved the classical astronomy in one text after another, but Middle Eastern research was limited to fine-tuning the great Ptolemaic system. As their observations called for tweaking the period or size of a planet's epicycle or deferent, so did they change their numbers. But even so, their scientific contributions were far superior to those of the Western world.

The most capable scientific mind in the Europe of the Dark Ages must have been Bede, known as the Venerable Bede, who lived from 672 to 735 near present-day Newcastle, in the very northern reaches of England up near the Scottish border. Here was perhaps the most knowledgeable scientist of the day living far from the intellectual centers of European society. It would be similar to today if the leading scientist lived in Whitehorse, capital of the Yukon Territory in northwestern Canada—although with email anyone's isolation nowadays would be much less than Bede's of that time. Bede was also a noteworthy historian, writing such books as the *Religious History of the English-Speaking People*. Bede somehow acquired a knowledge of the nature and causes of the tides and the increasing discrepancy between the Julian calendar, formulated and adopted by Julius Caesar in 46 B.C., and the date of Easter in the calendar of the Western Church. This date is made to fall on the sabbath following the first full Moon after the vernal equinox. The proper date for Easter could not be fixed to suit everyone, and this resulted in squabbling to the point of bloodshed in some western European lands. Bede knew that the Julian calendar was based on a year of exactly 365.25 days, a few minutes longer than the proper length of the year. Inserting an extra day every fourth year seemed to keep the calendar in proper alignment with the seasons. After the Nicaean Council in the year 325 fixed the date of the vernal equinox at March 21, the date of that event slipped backward by about 3 days every 400 years, so that by Bede's time it fell on or near March 17. Despite his protests nothing was done for another eight centuries, probably a record for procrastination, until Pope Gregory XIII proposed a new calendar to replace the old one in 1582, after Copernicus had changed the shape of the universe. At that time the disparity had built up to 10 days. Gregory and his astronomers realized that the year is but 365.2425 days in length, shorter than the Julian year by what amounted to 3 days every 400 years. He dropped 10 days from the Julian calendar to bring the vernal equinox back to March 21 and dropped 3 leap years every four centuries thereafter; thus 1700, 1800, and 1900 were leap years in the Julian but not in the Gregorian calendar. The years 1600 and 2000 were leap years in both but 2100, 2200, and 2300 will again be non–leap years in the Gregorian calendar. Even this calendar is slightly inaccurate since a more correct length of the year is 365.2422 days, shorter than the Gregorian approximation by 0.0003 days per year, or one day every 3,323 years or near; hence, sometime in the third or fourth millennium, yet another day should be dropped out by the people, if any, of that remote period. After all, by around A.D. 4909 the Gregorian calendar will become a whole day out of synch with the tropical calendar (the calendar that keeps the occurrences of the solstices and equinoxes on the same dates) we wish to promote, so people of that time will want to adjust it. But with no Moon, the whole calendar evolution, so dependent upon Easter in Christendom, and on the first sight of the crescent Moon in Islam, would have reshaped both religions into other forms.

Europe came alive intellectually some time around the twelfth and thirteenth centuries with the rise of universities and the scholastic life. For the first twelve centuries of the Christian era, Plato's dialogue *Timaeus* formed the basis of Western cosmology as much as any source. As Anthony Gottlieb explains in his *The Dream of Reason: A History of Philosophy from the Greeks to the Renaissance,* until the 1200s, when Latin translations of Aristotle and others first became available in the West, Plato's God of the *Timaeus* could be and was interpreted as the God of Genesis by the Christians. Gottlieb continues, "The main differences between Plato's God and the biblical one are these: his God is not the most important thing in the universe . . . he is not omnipotent but must co-operate with various natural forces." The Christian concept of God makes scientific inquiry unnecessary since divine purpose is not to be understood. But with Aristotle back in the hands of scholars, he was widely read and discussed, and later challenged.

While Dante and others of his time furthered the tradition of an Aristotelian universe complete with a heaven and a hell, scholars in England, France, and elsewhere were beginning to question any and all forms of geocentricity. They wondered why God's abode above the starry sphere had to whip around the sky every 24 hours at an unseemly speed. But if it was the Earth that did the rotating, how come a stone thrown directly upward did not land well to the west of the place where it was thrown? At least a handful of Westerners preceded Copernicus in the consideration, at least, of a Sun-centered model. One of the features that spurred scientists of the time to reexamine the Aristotelian model was the availability in the thirteenth century, for the first time, of translations of his work from Arabic, earlier translated from the original Greek directly into Latin, a language that many could read. This, along with the high quality of recent Arabic observations, led to better descriptions of the more advanced Ptolemaic system, numerically superior to Ptolemy's own, due to the fine-tuning that these observations provided. Such developments led directly to the first serious and widespread questioning of a stationary Earth, a century or two before Copernicus brought forth his fully reasoned alternative model. By that epoch we have the writings of Aristotle, Plato, and others in many languages translated directly from the original Greek, but at the time scholars had to be content with Latin translated from the Arabic tongues. If Aristotle had come upon a heliocentric system, the later Middle Ages would have been very different. Augustine might never have diminished us with his admonition: "There is another form of temptation, even more fraught with danger. This is the disease of curiosity. . . . It is this which drives us to try and discover the secrets of nature, those secrets which are beyond our understanding, which can avail us nothing which man should not wish to learn."

II
THE PLANETS

7

THE RINGS OF EARTH

October always brings on the frigid days, the days when the Sun proceeds south to the point where it passes behind our great celestial ring system. The average daily temperature has dropped by 10 degrees just this week and we must get out our winter wear. Then in early November, the warmth returns, at least for a while. Like Saturn, we have this giant ring system girdling the Earth; by day the rings seem about as thick as feathery cirrus clouds, too thin to appear gray or any color other than brilliant white or even to cast shadows. But they do dim the sunshine behind just enough to bring a break in our climate with this prewinter cold spell. By night up there in the sunshine they blind our attempts to pick out the fainter stars. Where they pass into the Earth's shadow, they are dark, even in the moonlight. If only we could get rid of them we could see the faint Milky Way all night every night, instead of only around midnight when the sunlit parts are at a minimum.

On clear nights when observatories are open to the public for a look through a telescope, astronomers try to show the Moon with its very visible mountains and craters along with the visible shadows they cast if the Moon is not full. Evenings with the Moon near the first quarter phase are most frequently selected for public nights because it is then high in the sky and rich in shadowy relief and because in the event of thin clouds, it is the brightest sight and therefore most likely to shine right through them.

After the Moon, the most popular sights are Jupiter and Saturn. Even at a modest magnifying power Jupiter shows its very oblate disk, wherein its equatorial diameter is fully 1 part in 15 larger than its polar diameter. For comparison, the oblateness of the Earth is but 1 part in 300; if we possessed the oblateness of Jupiter, our world would appear visibly flattened (see Fig. 7.1). Jupiter's great equatorial bulge owes its existence in large part to the fact that it spins around on its axis

Fig. 7.1 The Earth as it would appear with the oblateness of Jupiter. Photo: NASA.

once every 10 hours. With a 10-hour day, the massive Jupiter whirls so fast that such an ovate appearance is inevitable.

One can see belts of one type of cloud against others of a different color across the equatorial flank of this giant planet. Along one of the belts not far from Jupiter's equator lies the Great Red Spot, a feature known to exist since the invention of the telescope and maybe much longer. This giant vortex with circulation like a whirlpool or some kind of storm along the outer flanks of the planet is much larger than the entire Earth. This single hurricane, if it is one, would encompass all seven of our continents, the oceans, and much more. For 5 hours we can view the spot in even a modest telescope; then as it is carried by Jovian rotation around to the far side, it remains hidden for another 5 hours. But within a single night it can be seen at some hour. And finally there are the four big moons strung out along a line passing through the planet's wide equator, which forever changed Galileo's and our world when he first spotted them in 1610.

The one other perennial winner at the telescope is Saturn. With those rings Saturn can dominate any viewing session. A moon or two or three are generally visible and the disk of the planet appears like a smaller Jupiter (with an even greater degree of oblateness, 1 part in 10) with a less turbulent atmosphere, but it is the rings that amaze and entertain us. Mercury is a little thing lost in the twilight, and Venus and Mars are surprisingly almost featureless. Uranus and Neptune are visible as small greenish disks, so no other planet has a chance to attract attention as do Jupiter and Saturn, the two largest of our solar family. Star clusters, nebulae, and galaxies are faint, usually too faint to be seen against the light-polluted urban and suburban skies over most public observatories. That leaves the big three—the Moon, Jupiter, and Saturn—as the prime targets.

Saturn's rings are a marvel; over 170,000 miles across from one edge to the other, they are only a few miles thick at the very most and probably less. This means that twice in Saturn's 30-year orbital period about the Sun, we on the Earth see the rings edge-on in the scheme of its orientation to us, and they simply disappear altogether. The complex rings with gaps here and there cannot be other than collections of billions of tiny moonlets only a few feet or yards in size. They cannot be solid, like a giant phonograph record or CD, since no substance in the universe would be rigid enough to resist the tidal shear Saturn would impose. Saturn's gravity makes the closer moonlets orbit faster than the more distant ones, and solid rings would be torn apart at once. The rings came about either because a moon got too close to the planet and was torn into pieces by its gravitational field and tidal effect, or they represent original debris that never coalesced into a moon at all.

Since the start of space exploration, we have discovered that all three of the other major planets—Jupiter, Uranus, and Neptune—also have ring systems, but none are remotely as grand as Saturn's, and none can be seen directly from the Earth. All ring systems end up in the plane of the planet's equator and are paper-thin as a result. The little pieces of rock and ice undoubtedly collide with each other and split into smaller fragments.

What if the Earth had a ring system like Saturn's; what would we see? What if some of the original glop hung too close to our surface to withstand our gravitational divide and so could not form a respectable moon? Our impressions of the rings close up may derive more than anything else from the paintings of Chesley Bonestell (see Figs. 7.2 and 7.3). Half a century ago, on the eve of the Space Age, he painted many pictures depicting scenes as they might appear from the surfaces of other worlds. None were more impressive and memorable than those of views from the surfaces of Saturn and its biggest satellite, Titan.

Fig. 7.2 Saturn in the crescent phase as it would appear from the surface of its largest moon, Titan. Titan is now known to have a cloudy, relatively substantial atmosphere. Painting by Chesley Bonestell. Used with permission of Bonestell Space Art.

Bonestell's work has been felt by some to have enticed more young people into careers in astronomy and astronautics than any other stimulus.

From the Earth's equator any rings of ours might be scarcely visible at all since they would be so very thin, but as soon as one moved north or south, they would span the heavens, rising from the horizons in the east and the west to arch toward the zenith like a much, much more luminous Milky Way. They would shine all night every night with a ghostly white luminescence, forming a light much brighter than the full Moon and blotting out all but the brightest stars. In fact these brightest stars might well shine right through the tenuous rings, less than a mile in thickness.

A very recent theory contends that the Earth did indeed have a ring at one point in its history. About 35 million years ago this planet underwent a long period, perhaps as much as 100,000 years, during which the entire globe was unusually cold. One of the possibilities that brought this about involves a collision with an asteroid, but in this case it did not plow straight into the Earth as an earlier one did 65 million years ago, the one that finished off the dinosaurs. In this latter case the object sideswiped the Earth and splattered back out of the atmosphere into debris that spread out into a ring. The ring did not last for very long as pieces fell back into the atmosphere and onto the ground. This raises the point that the rings of Saturn and the other outer planets may also not have existed over their entire life spans. It is possible that whatever caused the formation of the rings in each case happened more recently.

If we look at Saturn we see that most of the time the rings appear incomplete; a gap appears between them and the planet. The rings are assuredly complete around the globe at all times, but parts of them lie in Saturn's shadow and can't be seen, having no light of their own. Our rings would also be dark in places as parts would pass through the Earth's shadow, but in the daytime they would appear complete and be easily seen. At dusk they would soar up from the western horizon past the high point toward the south, and as the evening wore on, the

Fig. 7.3 Rings similar to those of Saturn as they might appear from the Earth. Painting by Chesley Bonestell. Used with permission of Bonestell Space Art.

luminous part would recede toward the western horizon, reaching a minimum at midnight. Then they would appear to rise in the east until completeness at sunrise. We cannot describe the spectacle more succinctly than this unless we specify their diameter and extent. The Moon and planets move along or near the ecliptic inclined $23\frac{1}{2}°$ to the celestial equator, but the rings would lie along the equator, the extension of the equatorial plane projected onto the celestial sphere. One strategic difference should be pointed out between the ring system of Saturn and anything like that around here. Saturn's rings are made of ice—mostly water ice—with an admixture of rock. Out there water would be frozen under almost any conditions. But here near the Sun, any ring system could only be composed of rocky material similar to that of the Moon and the meteors. Ice is a good reflector of sunlight but rock is not. Bright as rings encircling our planet would be, they could not be expected to have the luster of those surrounding the outer planets, including Saturn. However, this would be more than offset by the fact that the Earth, being only a tenth as far from the Sun as Saturn, receives about a hundred times as much sunlight per unit area as that distant planet. Our rings would appear a good deal brighter to our eyes than their rings would appear to theirs.

Still and all, no fully dark night can ever be seen except from near the poles, where the rings, lying in the equatorial plane, are below the horizon at all times and are not visible. At lower latitudes, the sky is always too bright for most nighttime observing of faint objects, and stellar astronomy might well be retarded as a result. Our largest telescopes might be forced to hug northern Greenland and the Antarctic Plateau in order to observe the faint far-off galaxies and quasars on a dark night. The clouds and turbulence of the Arctic and Antarctic regions would make for a very attenuated observing schedule. More than ever would we need our Hubble telescope and others above the atmosphere, which can cut out the sky glow the rings would produce and evade them at times in order to view every part of the sky. In any case, we might still be doing last century's work even now. We might not yet have understood the thermonuclear processes in the interiors of the Sun and stars as early as we did. If that were the case, the Manhattan Project and World War II might have ended as they did because they depended on the fission of heavy elements, but we might not yet have developed the hydrogen bomb, which requires the fusion of hydrogen into helium just as takes place in the cores of stars.

8

NEXT DOOR TO
A GIANT

What a sight the Moon and nearby Jupiter make when they appear close to each other in the evening sky before and after sunset. The Moon subtends an angle of half a degree, 3 times that of Jupiter, if that huge planet were in the orbit occupied by Venus, or about 10 times the area in angular terms. Balanced against this is the fact that the big planet, perpetually enshrouded in a thick atmosphere, has 5 to 6 times the albedo or reflecting power of the airless Moon and, being only 70 percent of the distance of the Earth and Moon from the Sun, it receives twice as much sunlight per unit area of surface. Coupled together, these two factors, the greater sunlight plus the much larger albedo, offset the larger apparent area of the lunar disk. In the telescope we see two crescents, the larger Moon and the smaller, brighter Jupiter, the one revealing mountains and craters and the other, belts of clouds against a thick atmosphere, also of clouds but whiter. It is small wonder that, at moments like these, almost every observatory schedules an open house!

Jupiter, when appearing alongside the crescent Moon, shows a thicker crescent since the angle between the Sun and the Earth would be larger and closer to 90 degrees. But unlike the crescent Moon, the crescent Jupiter would not show the whole of its disk in faint light. We see the entire Moon faintly, an effect called the old Moon or earthshine, and first explained by Leonardo da Vinci. The Earth, which appears 40 times as bright as seen from the Moon as the Moon does to us, is easily bright enough to illuminate the lunar night side for us to see it. Jupiter has no bright Earth nearby and thus its dark side would remain invisible. Being as luminous as the Moon, Jupiter is always easily visible in broad daylight unless it is right up alongside the glare of the Sun.

If Jupiter happened to lie in the relatively close orbit of Venus, or even Mars, we would have one more object in our skies that would be seen as a globe with the naked eye. The four "terrestrial" or Earthlike planets—Mercury, Venus, Earth, and Mars—are similar to each other and not much else. They are all dense, some 4 to 5 times as dense as equivalent-sized blobs of water, whereas the Sun, the giant planets, and much of the rest of the solar system are about 1.5 times as dense as water. (Saturn is actually only about 0.8 times the density; the statement is frequently made that if there were an ocean large enough, Saturn could float on it.) The reason for this wide discrepancy in density between us and the Jovian worlds is very simple. The Sun and the rest of the universe are composed mostly of the lightest and simplest of all the elements, hydrogen. Hydrogen makes up over three-quarters of all the material of the universe. Every star, every nebula, every galaxy is composed mostly of hydrogen. It is by far the most abundant substance of all. Most of the rest is helium, the second lightest and simplest of the elements. Only about 3 percent of the universe, if that, is made up of the rest of the elements, the more than one hundred that spread from the third lightest, lithium, up through fermium and beyond. We must add here that some of the heavier elements may exist only in the laboratory. Astronomers have a very simple chemistry for most purposes; the amount of hydrogen is called X, that of helium is Y, and the rest combined are labeled Z such that $X + Y + Z =$ unity, where, as mentioned above, X amounts to over three-quarters of the entire lot by mass, Y accounts for most of the rest, and Z comes to only 3 percent at best. Within the scrambled mess we call Z, the most abundant elements are carbon, nitrogen, oxygen, neon, and one or two more that account for the bulk of it, all of which are among the lightest elements after helium. A few in the next tier of elements by weight—silicon, sulfur, aluminum, magnesium, and iron—make up the majority of all the rest. Beyond iron the elements are found in trace quantities at best; thorium, gold, lead, uranium, and the rest of that heavy group just don't amount to much.

What happened to the Earth and these other dense spheres? Where did the majority hydrogen and helium go in their case? Two reasons account for the shortfall, both due to the proximity of the Sun. Our inner system is hot, relatively speaking, compared to Jupiter and the outer solar system. The outermost terrestrial world, Mars, is only 1.5 AU from the Sun but Jupiter is 5 and Saturn about 10 AU away. The whole scale of the outer solar system is much larger than that of our inner planets. Water makes much of the difference; here in the inner solar system it is too warm to remain in its solid form as ice except in a very few areas, but from Jupiter on out, it can and does. And water, H_2O, is abundant out there and it is mostly hydrogen. The Sun imposes a second restriction on our hydrogen and helium; it acts gravitationally as a tidal force. Among the inner protoplanets, molecules of these two lightest and most buoyant gases, hydrogen and helium, were bounced up to the outer fringes of their atmospheres at the time of their formation and then got stripped away by the heat and the tidal influence of the nearby Sun. If, however, a planet the size of Jupiter were as close to the Sun as Venus, it is likely to have retained its initial hydrogen and helium—the velocity of escape from its great gravity would preclude much leakage into space by these lightest gases. We terrestrial planets didn't have a chance to retain much of them owing to their lightness and volatility, but we retained almost everything else, including oxygen. So here we are, dense worlds of iron, oxygen, and silicon, out of synch with the rest of the universe. The Moon is much the same as the Earthlike worlds— we can think of it astrophysically as a fifth dense planet, larger than Pluto—if it were in an orbit by itself around the Sun, it would be considered a planet in every sense.

So astronomers had it all worked out. Planets forming close to the Sun are small and dense whereas those much farther out can be large, composed mainly of hydrogen and helium like the Sun and the other stars, and could carry large satellite systems of ten or twenty or more at one time.

Then came the discoveries a few years ago of planets around other suns, nearby stars similar to our own Sun. And almost every one of them lay close to their primary stars and were massive—Jupiters in orbits supposedly reserved for the much smaller Mercurys and Venuses. Science does this to us from time to time; it can destroy the neatest of blueprints. And we can either accept the observations as truth or disregard them as the creationists do to preserve their Genesis.

Not all is yet lost. We cannot see even one of the dozens of Jupiters discovered circling other stars with any telescope. They are found indirectly by their gravitational perturbations of their central stars, as Jupiter would be seen to do to our Sun, much as a small dog dancing around on a leash would dislocate its owner at least a bit from the steady, linear walking motion the owner would have had without the perturbation of the dog. But the sample of Jupiters are at present biased by a large selection effect in favor of those among them that happen to be near their primary stars because the gravitational perturbations they impose on their central stars are larger and change faster. We won't discover the slower-moving Jupiters properly at or beyond our Jupiter's distance of 5 astronomical units until a few more years go by. Then we will see if these nearby Jupiters are rare anomalies or not.

Very recent observations now available have settled this matter. Of the more than one hundred planets presently known to circle other stars, all have masses of Jovian proportions or larger, and *none* remain in orbits close to their central stars, though they may swing into these perihelia upon occasion. Those that were seen to be closer to their central star than Mercury's distance to the Sun are now all known to move in highly eccentric orbits that carry them out to their proper distances where massive hydrogen-based planets can form and retain their large masses. Thus has the problem vanished as many supposed, and planetary astrophysicists can rest assured that their formation theories on this point, at least, remain viable.

Jupiter is 11 times the diameter of the Earth. In the orbit of Venus it would be large enough to be seen as a disk, smaller than the Sun or the Moon, but showing the full set of phases to the naked eye. Jupiter's phases, being visible to our unaided gaze, vitiate the models of Aristotle and Ptolemy because it would be obvious that Jupiter goes around the Sun and not the Earth even to them. When on our side of the Sun, Jupiter appears relatively large because it is closer to us, and can be seen as a crescent. Later, when it has swung to the other side of the Sun, it appears in the gibbous phase, but smaller in angular size, and in some cases at its farthest, it is too small to still present a disk to the naked eye. This is exactly what Venus does, but Venus is too small to be seen as a disk by the unaided eye at any time and had to await Galileo and his telescopes to reveal its various phases.

One difference between the phases of Venus (or in our imagined case, Jupiter) and those of the Moon becomes immediately apparent. The Moon revolves about the Earth in a near circular orbit with a 5 percent eccentricity. Thus its apparent size in the sky changes very little, and the change goes unnoticed by the casual observer. But Venus varies from about 25 to 160 million miles away as it passes in front and behind the Sun. At the closest it appears some 6 times as large as it does at its greatest distance; this would be so evident that the Earth-centered systems from Aristotle to Ptolemy could not have been postulated at all, for they would have *seen* that this Jupiter goes around the Sun. An intermediate arrangement like that advanced by Tycho Brahe just after Copernicus would be possible and yet retain the geocentric principle. Tycho's was a compromise, sociopolitical as well as astronomical. He proposed that the planets did orbit the Sun while it, carrying them along like satellites, swung around the Earth. The Earth did not move, according to Tycho, but it was not the center of all motion either. In every aspect except the fixation of the center, Tycho's scheme is identical to that of Copernicus. It preserved only the geocentricity favored by the Church. Just which cosmology would have dominated in the Middle Ages in the event of a disklike Jupiter in the sky is unknown, but the Ptolemaic model is not likely to have survived for long.

Jupiter in the orbit of Mars would pose a stickier problem for the pre-Copernican astronomers. Large enough to appear as a disk when closest, it would shrink to a point when more distant. But worse for any possibility of a compromise system, the apparent motions of planets beyond the Earth—from Mars through Pluto—show very complex motions. Each of them appears at times to reverse its normal eastward motion among the stars and move in a retrograde (westerly) direction for a while before resuming again its normal eastward progress. In an Earth-centered cosmos this is nearly impossible to explain; one needs the complicated gimmickry of Ptolemy with epicycles and deferents of just the right size to match the observations of planetary positions and motions. The Sun-centered arrangement of Copernicus, remodeled by Kepler, made it far simpler to explain this stop-start retrograde movement. Not only did Kepler's model subscribe to the principle of Ockham's Razor (the simpler explanation is to be preferred over those of greater complexity, other features being equal), but to Kepler it made physical sense when he found that orbits are elliptical, not circular.

Another consequence of a nearby Jupiter would be the enormous gravitational and tidal perturbations this big world, 318 times as massive as our own, would impose on the Earth-Moon system. Our two orbits would be pulled and hauled around to the extent that climatic influences such as the long-term Milankovic effects discussed in Chapter 4 might be significantly altered—ice ages might be longer or shorter, more or less frequent, or absent altogether. Tidal tables might need substantial revision since Jupiter would then become a major player in the tidal effect on our oceans and atmosphere.

In any case, with one of the five naked-eye planets appearing at times as a disk, astronomical history would not have been the same. From Hellenistic times at least, the Alexandrian astronomers would probably have concurred with the renegade Aristarchus and placed the Sun at the center of planetary motion à la Tycho Brahe, if not of the rest of their starry universe. The Islamic scholars would, more credibly than not, have retained this concept, and even the Western Church might have allowed the seeming heresy throughout the Dark Ages. The effect on the lives of Copernicus, Galileo, and above all the unfortunate Giordano Bruno, burned at the stake for heresy, are incalculable, possibly to the extent that only Galileo of the three would even be known today, because he might still be the first to use a telescope astronomically, and he was also the father of experimental physics.

There are reasons why, of all the planets and satellites, Venus is the world that most closely resembles the Earth in the large overall properties, among them size, mass, gravitation, and average density. It is only 10 percent smaller in diameter than is our world and 20 percent smaller in mass, and as a result of this near equivalence the planet possesses a molten core and a seismic interior very like ours, replete with volcanism, venusquakes, and continental drift. Larger planets stand to the rear in our solar system. Jupiter and the other giants were thought to form away from the Sun, where the solar heat and gravitation are much weaker and a planet can keep all of its original mass. Then we discovered planets of Jovian mass closer to their central stars than even Mercury is to the Sun. These things happen in science from time to time; science is a self-correcting activity, unlike most others, and we will adapt our theories again to the new observations.

But with size and mass the similarity ends; at the surface Venus is nothing like the Earth. Through a small telescope Venus is lovely but with an exterior that is bright and nearly featureless in any of its phases. To the eye it is seen as our brilliant evening star when east of the Sun, where it can be observed in the western sky before and after sunset, and it is our morning star before or at sunrise, when it has moved to the west of the Sun.

In many cultures, occidental and oriental, Venus has been associated with love and beauty. Totally enshrouded in clouds, its unique brilliance and pristine whiteness make this association a natural one. Mars, in contrast, is red above all, a rusty mix of desert sand and rock. Although not exactly blood red, Mars is frequently identified as the god of war. Contrast the first two sections or tableaux of *The Planets,* the famous orchestral suite by Gustav Holst. Holst felt an astrological communion with the solar system, and his work emphasizes the connection. The first section is devoted to Mars, the Bringer of War, and builds to an incredible intensity on a single, pounding rhythmic theme. This is followed by Venus, the Bringer of Peace, with an appropriately soft, flowing quality in total contradistinction to the music representative of Mars.

Somehow the two images have been integrated and blended with these two most Earthlike planets to such an extent that the recent discoveries as to their surface conditions came as a surprise. As ground-based astronomy was unable to pierce the atmospheric veil surrounding Venus, thus failing to discover its surface temperatures, atmospheric content, and even its period of rotation, we began to conceive it as our own imagined warmer version of the Earth. A century ago it was popular to portray that world as a planet-wide tropical rain forest populated with dinosaurs and other creatures generally similar to those of our own warm Mesozoic Era.

Mars, on the other hand, was seen as a desolate, almost sterile globe with an intelligent race far older and more technically advanced than we are, though possibly extinct. The Boston Brahmin and erstwhile astronomer Percival Lowell (one of the Lowells, who spoke only to Cabots, who spoke only to God) and others thought they saw straight lines crossing the reddish disk and imagined an intelligent race irrigating their vanishing forests and crops with canals transporting water from the snow-white polar caps to the equatorial regions. H. G. Wells traded heavily on this image with his *War of the Worlds* of 1898, in which these Martians attempted to fulfill their dream of invading and settling the much more verdant Earth. The biota of Earth won the war but not until much mischief and damage had been done. To cement Mars forever in our minds as a world of horror, death, and little green men, young Orson Welles terrified half the population with a radio simulation of newscasts outlining H. G. Wells's story in 1930s modern dress. Similar outbursts of panic occurred in Latin America when a Spanish version was broadcast at a later date.

Only the images put out by the Viking and other Mars landers over the last 30 years have dispelled this ominous conception. The real Martian surface is hostile indeed but much less so than any other world but our own. With just under 1 percent of our atmosphere and its surface pressure, and 95 percent carbon dioxide at that, it would impose an air pressure found here about 120,000 feet above the ground. Temperatures could reach livable conditions at noon in this thin air near the equator but fall to more than 100 degrees below zero at night away from its tropics. Still, if properly prepared and suited astronauts can survive on the Moon, they can do so on Mars as well.

Far more than either the Moon or Mars, does Venus resemble Hell.

Like Mars, Venus has an atmosphere that is about 95 percent carbon dioxide. Carbon dioxide is a greenhouse gas, the most prevalent one in our own atmosphere (after water vapor), where its concentration is less than 0.04 percent of the total. But like methane, water vapor, ozone, and other greenhouse gases, all with three or more atoms in each molecule, carbon dioxide is transparent or at least translucent to the incoming solar radiation, mostly in the visual region of the electromagnetic spectrum, but opaque to the reradiation in the infrared region of the spectrum back into space from the planet's surface. It influences our climatic regime far out of all proportion to its trace abundance. In this manner it is like the glass of a greenhouse, which by doing the same thing can block the reradiation and create a tropical climate in wintry areas for plants inside. More common than a greenhouse is an automobile in the summer sun with its windows closed. Woe betide the person who leaves a child or a pet for any length of time in such a circumstance, for the conditions are soon fatal, with the temperature inside reaching 140 degrees or more very quickly.

On the Earth, alone among the planets, the most important of our greenhouse gases is water vapor. This gas alone raises the average surface temperature here by about 120 degrees over and above the temperature if we had no natural greenhouse effect at all. Even the small quantity of carbon dioxide in the Earth's atmosphere raises the temperature we would have without this gas by a few degrees. The Earth is not without carbon dioxide, not at all. Twenty thousand times the amount of this gas in the air exists in our rocks and elsewhere. We are very fortunate that it is there, and we would do well to keep it there. On Venus the 95 percent abundance of carbon dioxide in an atmosphere 90 times our own in quantity and surface pressure makes for a prodigious lot of carbon dioxide and creates what is known as a runaway greenhouse effect. Venus has as much air as the Earth would have if all of our oceans boiled off into steam. The surface temperature on Venus is not just moderately hotter, which it would be just from the sunlight it receives, but hundreds of degrees hotter, being almost 900 degrees Fahrenheit at its surface. With its great heat and its carbon dioxide with a sprinkling of sulfuric acid in its air for good measure, Venus is rather more hostile to life than the basement of a department store during a sale on a hot summer day.

In another "what if," let us suppose that Venus and Mars changed orbits. Now the order from the Sun stands: Mercury, Mars, Earth, Venus.

Oddly enough, such an exchange would make both planets more habitable than they are now, more like the Earth at their surfaces and in their atmospheres. Mars, only a little more than half our world's diameter, would not have the bulk to promote an atmosphere as dense and, even if mostly carbon dioxide, could not pump up such a runaway greenhouse effect as Venus has done. Temperatures and pressures would be substantially closer to ours; how much closer might require the full history of conditions on the then maybe not so red planet.

Venus, out there in colder circumstances, could not have mustered the thick blanket of air it now has. Most of that air would have condensed out into something else, perhaps water ice if a mechanism there acted as it did on the Earth to take up the carbon dioxide into water and rock and, above all, plants. Were this to fail, much of the carbon dioxide would have frozen directly into dry ice, as it is called in its solid form, especially in the polar areas. Whatever gas remained would be much less than on the Venus of today, and certainly at a much lower temperature, somewhere between the thermal boundaries of the present Earth and the present Mars, more livable by far than its present Hell.

9

DOUBLE PLANET

I looked upon the Martian surface. From a slight rise I could look straight along one of the "canals" that covered the planet when an intelligent race was assumed to have mounted a great effort to irrigate the equatorial regions from the melting polar caps during each hemisphere's springtime. The Sun shone along the canal low in the deep southwestern sky; it would set in just a couple of hours. The borders of the canal were not even, but rough, with edges that just petered out onto the desert sands. Any apparent oasis was limited to the canal itself. In fact, from this small mount I could see that the canal's linearity was not entirely convincing and might have been the result of a chance alignment of natural oases. We had known of their illusory nature, these unreal chimeras of seeming linearity, since the first unmanned Mariner space probes of decades ago closely scanned the planet's surface.

All my life I had dreamed of visiting Mars; it had been my ambition since childhood. How excited I was when my name was announced as being among the crew of our first manned mission to the red planet. As one of five selected to participate in the venture, I thrilled at the heavy doses of training and exposure to the conditions to which we would be subject for the months we were to spend on the planet's surface before the return to Earth. I lived eagerly for those days, and now how abhorrent it was. As the mission's only survivor of that fateful landing and with no chance of getting home again, I was the red world's only inhabitant. I had abundant supplies of food, water, and air, intended for five, and would endure here for months at the very least. But the compact library of reading material and recorded music was, despite its treasures, no substitute for the warm companionship of my fellow crewmembers.

I never before felt so utterly and implacably alone. Where had the Martian people all gone if indeed they were ever here? Was this world once alive with the gaiety of the populace, the cries of children at play? Now only the noisy clamor of the ever present gusty wind, sometimes swirling up into a gale-force blow and whisking the sand into tiny eddies of activity, broke the stillness of this lifeless world.

The Sun was smaller in these Martian skies, not much more than half the size it appears from Earth—my home, now beyond reach. The sky itself looked deceptively dark blue, darker even than it appears in some of Vincent van Gogh's blue and mustard last paintings. A few light cirrus clouds here and there dotted the scene, but they were not thick enough to blot out the brightest stars that were only just visible during the daylight hours in the thin evening air.

It came to my mind then, the doleful mourning of "The Swan of Tuonela." This is one of four orchestral tone poems by Jean Sibelius, each a depiction of a scene from the Kalevala, the great Finnish national epic, and surely the loneliest music ever composed. It reveals the isolation, the utterly hopeless cancer of the soul that can disable one's sense of the future. I felt that life had ended for me in all ways but the final one.

Even the cheery sight of the triple planet in the sky not far above the Sun only reminded me of the life I had left behind. There was Venus, a creamy white point not much less brilliant than we see it from our Earth as the evening star. But then not far above it shone a brighter blue-tinted world with a white star just alongside it. Our fair Earth and Moon formed a brilliant double planet aloft in the Martian daylight sky. Now what would the ancients—Aristotle and Hipparchus and Ptolemy—have made of that?

Tragic overtones aside, this is a sight that will greet the first hominids to visit Mars. What would they make of this tableau, those who promulgated a cosmos requiring the Earth to be the center of all motion, a point considered but discarded by the Greeks and imposed on the rest of us by the Church for centuries?

The Earth and the Moon form a nearly unique pair. They are closer to a double planet than any other except for Pluto and its moon. Jupiter, Saturn, and Neptune also have large moons (in Jupiter's case, four of them), but the planets are so much larger that they dwarf the moons that appear only as tiny stars to us. No moon but ours is visible without binoculars or a telescope from another planet, but if any of our closer terrestrial neighbors sported one as large as our Moon, we could easily see it in our sky. Perceived from Venus, our Earth at its closest and brightest appears bright and blue at magnitude -6, several times as luminous as Venus, the brightest object in our skies after the Sun and Moon. Even the Moon seen from Venus when Venus is at its closest, about 26 million miles away, would outshine all but the Sun and the

Earth, being about as bright as Jupiter, Mars, or Mercury at their best, at about magnitude -2. At new Moon, the Earth-Moon system would be unresolved to the eye, appearing as one object. But then over the next seven days the Moon would appear to draw farther and farther from the more lustrous Earth until it stood about one-half degree away (to the right from the northern hemisphere of Venus), about the angular diameter of the Sun or the Moon as they are viewed by us. In the following two weeks the Moon would appear to approach the brighter Earth and pass on to the left side to an equal distance. In the fourth week the Moon would again sidle up to the Earth. The blue Earth and yellow-white Moon form a stunning sight, and if Venus were not perpetually overcast and intelligent beings lived there, they would certainly dream of a flight to Earth, as we do about one to Mars.

From Mars the pair appears with about half the separation or $\frac{1}{4}°$ at its widest and about three magnitudes fainter than from Venus, and from the outer planets they couldn't be seen at all without optical aid. It would be obvious that the fainter satellite goes around the brighter planet (Venus or Mars) and nothing else. Our world could not be accepted as the center of all motion as it has been. It is not possible to tell whether the Copernican model would have been accepted in classical times, but it is a tantalizing thought. Might Aristotle have divined that the Earth-Moon system resembled that of either nearby planet with a visible moon, thus gaining two thousand years of foresight on one point at least? Might astrology and the Church in medieval times each have formed along different lines and, if so, how? We cannot tell, but the sight of a moon (or two moons?) circling the radiant evening star would make for a mini–solar system that could not be ignored. Much of the speculation of the previous chapter applies here as well.

Just what did happen in the real case of post-Renaissance cosmology with Venus and Jupiter where we know them to be? Nicolaus Copernicus published his book, *De Revolutionibus Orbium Coelestum (On the Revolutions of the Heavenly Spheres)*, in 1543, the year of his death. For the next century, his system and the Earth-centered system of the ancients stood side by side, their merits and shortcomings discussed and debated by many. Tycho's compromise arrangement entered the debate as well. Although the proper center of the solar system occupied center stage, other points also differed between them. The most sophisticated of the geocentric theories was proposed by Ptolemy about A.D. 150 in his book now known as the *Almagest,* but the Church and others through the Middle Ages gradually replaced it with the simpler one of Aristotle. Listed here are some of the points of difference between the establishment interpretation and the heretical views of Kepler, Galileo, and a number of other astronomers of the late sixteenth and seventeenth centuries.

Orthodoxy, Established Dogma	Heresy
Spherical Earth	No difference
Stationary Earth	Rotation and revolution about the Sun
Earth at center of solar system	Sun at center
Circular motion at constant speed	Elliptical motion at variable speed
Dualistic nature of Heaven and Earth	Universal laws of physics and chemistry
Immutability, changelessness over time	Changes may occur
Plenitude, no gaps or vacuum	Mostly empty space
Stars just beyond Saturn	Stars as suns and very distant
Earth center of all motion	Other planets may have satellites

A round Earth was widely accepted in Greece by about 500 B.C., but a stationary Earth at the center, as opposed to one that spun on its axis and revolved around the Sun, was the principal point of contention. That all motion must be in the form of a circle was not disproven until Kepler did so about 1609, over the opposition even of Galileo, with whom he corresponded. Aristotle had particularly stressed the fundamental difference between this lowly globe made of earth, air, fire, and water, and Heaven from the Moon on out, which he thought to be comprised of the quintessence, a fifth element of crystalline purity that filled all of celestial space with no gaps anywhere. From Copernicus through Newton, whose work was first published in 1687, this stance weakened as it became ever more obvious that the laws of motion and gravitation are universal.

In 1572 a supernova burst forth in Cassiopeia; it became as bright as Venus and was visible in the daytime for some weeks until it gradually diminished in luminosity. Though noticed by everyone, it was carefully observed by Tycho and is still known as Tycho's Nova. He observed that it remained stationary with respect to the other stars, sharing only their motion caused by our rotation, and he concluded that the supernova belonged to and in the heavens and not in our atmosphere. No one could ignore a star of this magnitude, and those who maintained the immutability of the celestial realm in this instance, at least, were shown to be wrong. In 1610 Galileo discovered four little worlds that certainly circled around Jupiter, resembling a miniature solar system. Clearly, our planet was not the center of all motion.

As more and more of the Aristotelian precepts were proven incorrect, the geocentric system gradually crumbled. The burning at the stake of Giordano Bruno in Rome in 1600 (for heretical views including the Copernican theory) and the unpleasantness between Galileo and the Church publicized the subject into world prominence. The only surviving point in Aristotle's favor was that no one had detected the elusive parallactic motion that stars must show if we spin about the Sun. Failure after failure to observe it did not prevent the ultimate acceptance of supremely large stellar distances over the alternative view that the Earth did not move, or of the correct explanation—that the stars had to be suns far beyond any of the planets, too far for their motion to be observed. The retreat of scripture on one point after another meant that the whole biblical scheme of the architecture of the universe became untenable two centuries before the first parallaxes were measured in 1838.

One subtle point involving the distances of the stars did favor the Copernicans, and this was the appearance of the constellations throughout the year. Picture the location of Orion, for example; in the wintertime this bright star group is about opposite the Sun in the sky, and at that time we are closest to the stars in Orion, however far away they may be. Several months later it is on nearly the same side as the Sun in the late springtime, not long before it is lost altogether in the solar glare. Orion is almost two astronomical units farther from us in late spring than in the winter and must appear smaller in the sky. If the stars were affixed to a sphere not far beyond Saturn, their distance from the Earth would vary by an amount of 2 AU divided by the diameter of the sphere holding the stars. If this diameter were to be 20 or 40 or even 100 AU, the repeated expansion and contraction of Orion would have been easily detectable by the ancients—but it was not. It is not hundreds, but millions of times as far from us as is the planetary reign, and over that span changes in the angles between Orion's stars could not be detected; in fact they could only barely be measured even today, when we know full well what the amount of change must be. This apparent constancy of size of star patterns places the stars far out beyond all of the planets and points to the second and correct reason for the apparent absence of stellar parallactic motion.

As noted above, with a moon visible to the naked eye circling a nearby planet, Venus or Mars, Aristotle's and Ptolemy's cosmologies might have been very different. Seeing a bright moon circle either planet informs us that the Earth is not the center of all motion. The same conclusion could be drawn if our Moon had a smaller but still visible satellite in orbit around it.

Two other features of our universe were first brought to attention by the telescope. Galileo immediately noticed that with each successive improvement in the power of his telescopes, he saw ever more stars. Stars fainter than the sixth magnitude, the serviceable limiting magnitude of the faintest naked-eye star, kept springing up wherever he looked. He saw stars of magnitudes seven, eight, and so on. This raised troublesome points; just what did God make all these stars for? No one could have seen them, so what was their purpose? Secondly, Galileo and his successors promptly made many improvements in optics that resulted in ever larger telescopes of higher optical quality. With each advance in magnifying and light-gathering power, yet fainter stars could be seen, and the planets kept appearing larger, but the stars, other than the Sun, remained the same pointlike entities they had always appeared to be. Even today we cannot see stars as disks in our largest instruments, at least not directly. We have indirect techniques for measuring stellar diameters, such as interferometry, but viewed directly through a telescope, stars remain points of light. This led to an acceptance that stars are materially very different from planets and unlike them. Stars are incredibly far off, so much so that they must be as bright as the Sun. The Sun was recognized for the first time for what it is—a star!

10

DEBRIS IN
THE SOLAR SYSTEM

Halley's comet is the only bright, spectacular comet that has a period of less than a few thousand years. What if its period were not about 76 years but only 5 or 10? Many comets move about that quickly, with Encke's comet orbiting in only 3.3 years, the briefest of all. But, like the majority of comets, none of these are visible to the naked eye. They once had much longer periods but perturbations by the major planets, Jupiter in particular, disturbed them to the degree that their periods were vastly shortened. In fact, one comet, Shoemaker-Levy 9, was captured by Jupiter and proceeded to circle around it until its spectacular collision with that big world in 1994.

Comets move in long, cigar-shaped elliptical orbits. As an example, Halley's comet migrates around in a highly elliptical orbit of eccentricity 0.967. This means that at its closest point to the Sun, or perihelion, it is but 0.6 AU from the Sun, closer even than Venus at 0.7. But at aphelion, the farthest point, half an orbit or some 38 years later, it sails out to over 35 AU, well beyond the orbit of Neptune. At that time it can only be detected using one of the very largest telescopes with advanced ancillary detection equipment.

If Halley's comet had been forced into a much smaller orbit, it wouldn't be called Halley's comet. Edmond Halley saw it in 1682 along with his older colleague, Sir Isaac Newton, whose laws the two of them used to show that every 76 years a bright comet appeared in the same place. They concluded that the earlier appearances in 1531 and 1607 were of the same object they themselves observed. (We note here that because comets are of very small mass, they get jerked around by the massive major planets. And because they shed some of their own mass whenever they approach the Sun, their orbital periods can change, in Halley's case from 75 to nearly 78 years.) Until that time comets were thought to belong to the mundane world, and considered to be apparitions in our atmosphere. Halley made a prediction that the comet would return again in 1758 and indicated where in the sky it would first appear. It did just that and the world never again seriously considered scientific laws to be other than universal, an inestimably great event in human science and thought. Other comets, including Encke's, cited above, are named after their discoverers, but Halley's is groundbreaking here too. Before his time, a comet was known by the year of its discovery, thus the Great Comet of 1682, for example, but when he convinced the world that this comet, at least, was periodic, it was properly named in his honor, the first named and still much the most famous.

More frequent appearances would very probably have been noticed by many of the scientists of classical times and assigned then to the realm of the planets, and not to our mundane atmosphere. It is hard to imagine that the ancients would not have realized that it moved in a very elongated orbit, like all comets, and that circular motion did not hold for everything. Ptolemy might well have rejected circles in favor of ellipses, 1,500 years before Kepler did so.

Only a small minority of Earth's inhabitants has lived through two separate and successive apparitions of Halley's comet. Mark Twain did, for he was born in 1835 when the comet flared into its magnificent form, complete with tail, for a few months, and on its next visit in 1910, he died. Anticipating the coincidence, he remarked, "I came in with Halley's Comet in 1835. It is coming again next year (1910) and I expect to go out with it. It will be the greatest disappointment of my life if I don't go out with Halley's Comet. The Almighty has said, no doubt: '[N]ow here are these two unaccountable freaks; they came in together, they must go out together.'" I have known only a few people who saw and remembered its 1910 visit and were alive to see it again in 1986. Among those who first saw it in that latter year, a few of the younger among us will live to see its next return in 2061 if they can guess which are really the best foods to eat to ensure a long life. But if it returned every five years or so, the count of visits observed could be well over a dozen and into the teens or even higher over a long lifetime. It would then be almost as frequent an occurrence as the Olympics. Surely Aristotle and his followers would have accepted its true position in the cosmos, if not its physical nature. With Halley's in place, would we not conclude rightly that other brilliant comets, one of which comes along every few years, would also be relegated to the heavenly domain? The great comet of 1997, named Comet Hale-Bopp after its two discoverers, was probably seen by more millions of people than any other. It seemed to settle into a very favorable spot in the celestial sphere for convenient evening viewing, night after night at dusk for over a month, and being brighter than all but one or two stars, it shone even through skies in suburbia affected by light pollution, though probably not in city centers, where the airglow renders the sky just too bright.

Comets form one of the two major groups of solar system debris; the other is that of the asteroids. Each group is characterized by its location within the system and by its general chemical and physical makeup. Most asteroids are found in orbits lying between those of Mars and Jupiter. In 1766 a law of planetary distances was formulated by two German astronomers, Johann Bode and Johann Titius. After many years the Bode-Titius law was found to be without physical reality, but it is a convenience in recalling the distances of the planets from the Sun, with some large-scale deviations along the way. This scheme predicts the distances shown in Table 2. It calls for doubling the number in the first column for each successive planet and adding a constant of 0.4 astronomical units for a predicted distance given in the fourth column. The true distance for each is placed for comparison in the fifth or penultimate column, and the periods of their orbits are shown in the last. The law holds remarkably well from Venus through Uranus. It violates its own rule for Mercury in that the first entry should be 0.15, not 0.0, in order that the value for Venus is properly double that for Mercury. If we overlook the discrepancy for Mercury, we get an amazing prediction for all of the naked-eye planets, except for a gap between Mars and Jupiter; thus when Uranus was discovered in 1781 and Ceres in 1801, the law was seemingly confirmed, since they both fit their predicted distances closely. The later discoveries of Neptune in 1846 and Pluto in 1930 rendered the law into nothing more than a convenient mnemonic. The scale of the outermost planets is nothing like our own; Neptune, for example, will not have completed even one turn around the Sun since its discovery until 2010, still in our future, and Pluto will not have made the trip until about the year 2177.

The discovery of Ceres was the more remarkable of the two milestones. It was the first asteroid to be discovered and is still much the largest among them; furthermore, it so nicely filled in the obvious gap between Mars and Jupiter.

But Ceres turned out to be a little thing, now known to be only 578 miles in diameter; the Moon is much larger at 2,160 miles in diameter. Two of the next three asteroids to be discovered, Pallas and

TABLE 2. THE BODE-TITIUS LAW

Planet					Distance	Period
Mercury	0.0*	+	0.4	=	0.4*	0.4	88 days
Venus	0.3	+	0.4	=	0.7	0.7	225
Earth	0.6	+	0.4	=	1.0	1.0	365
Mars	1.2	+	0.4	=	1.6	1.5	687
Ceres	2.4	+	0.4	=	2.8	2.8	4.6 years
Jupiter	4.8	+	0.4	=	5.2	5.2	11.9
Saturn	9.6	+	0.4	=	10.0	9.5	29.7
Uranus	19.2	+	0.4	=	19.6	19.2	84
Neptune	38.4	+	0.4	=	38.8	30.1	164
Pluto	76.8	+	0.4	=	77.2	39.4	247

*The rule ideally should be converted to predict a distance for Mercury not of 0.4 but 0.55 AU.

Vesta, are the next largest after Ceres, about half its diameter (around 300 miles across) and thus about one-eighth its volume and mass. Only 16 are now known that span even half that smaller diameter, or some 150 miles in size. By now we have probably found all of that size, but asteroids range down to ever smaller sizes and become more numerous as the diameter decreases. Thus we know 99 percent of all the ones over 100 miles, but between 10 and 100 miles, we know only something like half the total number in the solar system. We know enough, however, to estimate closely the size of the object that would result from a combination of all asteroids into one "superasteroid," if that term can be used for a still small bit of debris. Such a body would be only about 932 miles across, according to NASA, still far smaller than the Moon or even Pluto. It is very likely that the swarm of these minor planets resulted from their inability to coalesce into one large one under the domineering gravitational influence of nearby Jupiter.

The visibility of this superasteroid would be disappointing. At its brightest, Ceres is at about the seventh magnitude, nicely seen in 7×50 binoculars (the binoculars of choice for the night sky) but not by the naked eye. Vesta, one of the two next largest, is actually a bit brighter than Ceres in our sky. This is because Vesta's albedo is two to three times higher than that of Ceres, and it revolves a little closer to the Sun. Ceres and Pallas belong to the majority of asteroid types, the C class, which is formed of carbonaceous chondrites with some silicon and metals such as iron and nickel in the mix. Vesta is basaltic and may have been subject to extreme heat at some point in its past. If all the asteroids were lumped together into one with a diameter of 932 miles, it would still not be bright enough to be seen even on the best of nights except for the sharp-eyed among us. It would appear one magnitude brighter than Ceres, or about sixth magnitude at best, about as bright as Uranus, which, at 5.7, can just be seen with the eye under the best conditions.

The great majority of these minor planets hover near the distance of Ceres from the Sun, within 0.4 of its 2.8 astronomical units, but a tiny few are known to approach the orbit of the Earth from time to time. These are the so-called Earth-crossers, the ones that could at some point career into our world with devastating effect. We know of one such calamity for certain; about 65 million years ago an asteroid, or possibly a comet, perhaps 6 to 10 miles across, crashed into the Yucatan Peninsula in Mexico and tore our world apart. A majority of all species of life promptly went extinct, including all of the dinosaurs. Imagine a mountain hurtling into the Earth at about 15 or 20 miles per *second!* The energy released far exceeded that of all the nuclear arsenals of the world detonated together.

The impact shattered and vaporized the intruder at once as it burrowed into the ground a distance of several times its own size. It caused fiery stuff to ignite forest fires in all corners of the world. Then the dust, soil, and detritus thrown aloft into the air quickly imposed frigid conditions around the globe over weeks or months, blocking out a significant portion of the incoming sunlight and chilling the world into an extreme version of what Carl Sagan and others labeled a "nuclear winter," freezing and starving many creatures and sending whole food chains into rapid extinction.

This calamity is called the K/T event, after the sudden end it imposed on the Cretaceous (spelled with a K in German) Period, the last of the three divisions of the Mesozoic Era, the age of dinosaurs and reptiles, delineating it sharply from the beginning of the Tertiary Period, the so-called age of mammals, now 65 million years old. It is the last of five known "mass extinctions," in which most of the world's extant species simply vanished. Earlier mass extinctions occurring about 250 and 205 million years in the past may have had volcanic origins, but we know this last one to be due mostly to the shattering impact from space. The earlier ones are called the T/J and P/T events.

If truly all the asteroids were combined into a larger minor planet moving in the well-behaved orbit of Ceres, 2.8 times Earth's distance from the Sun, we would not need to fear a repeat of the K/T disaster, as we do now; the probability of a strike is very low in any one century, though inevitable in the far longer run. Our proper concern with even a small possibility of the event has sparked great public interest and spawned many books and at least two recent disaster movies. But *Homo sapiens* is close to developing the technology to prevent any future doomsdays of this magnitude. If we are lucky and smart, before another misplaced asteroid or comet comes along, we will have the means to deter it into an orbit that misses the Earth.

Of course, if there had been no loose asteroids or comets to hector our planet and its inhabitants, there would have been no fiery end to the Mesozoic Era, and the dominant life form, the dinosaur, might have developed into an intelligent species as portrayed in the recent super-spectacular film *Jurassic Park,* in which a race of saurian velociraptors might have governed the world from that time to this and prevented the subsequent rise of mammalian intelligence.

A large comet could also throw our world into a maelstrom and end human civilization. We know that comets are born in the third of three belts of debris in the solar system. The first is the asteroid belt between Mars and Jupiter, whereas the second stretches out beyond the orbit of Neptune at 30 astronomical units from the Sun, and includes Pluto and its moon among its members. For dynamical reasons it ends near 50 AU with a possible few venturing just beyond this limit. This is known as the Kuiper belt after its discoverer, or (sometimes) the trans-Neptunian belt, filled with icy worlds of which Pluto may just not be the largest.

The third belt extends far beyond Pluto and includes millions of comets in what is known as the Oort cloud, similarly named after its discoverer. Out there, at about 10,000 to 50,000 AU from the Earth and the Sun, these small bodies swirl in orbits with periods lasting many thousands of years. Once in a while one of the comets from this distant Oort cloud is induced into an orbital change that throws it into the inner part of the system. In those far reaches comets are icy worlds, mostly smaller than 10 miles across, with admixtures of small particles of rock. Think of them as dirty snowballs. Whenever one passes closer to the Sun than Jupiter and into the domain of our terrestrial planets, the increased solar radiation brings about a sublimation of the major icy part. This causes particles of gas and dust to stream out of the comet proper and form the giant tail that makes for such an impressive show.

Then when the comet retreats back to Jupiter's distance, it will lose the tail and appear again as a small dirty snowball. But in the course of this close pass to the Sun, the comet loses some of its mass in the form of the emitted gases and rocky particles. The closer a comet approaches the Sun, the greater the portion of its total bulk that will be lost in the flyby. A few stray so close that they are torn to pieces by the solar heat and tidal effects and are never seen as comets again. In such a case the snowball is gone but the dirt in it continues on in the comet's orbit. These tiny objects, most smaller than a grain of sand, are seen later only if they enter our atmosphere. Then the friction encountered burns up the particle, causing it to glow for a few seconds or less, and we see it over a lake or beach of a summer night as a meteor or shooting star. Gradually over the eons, the meteors sailing along in the old comet's orbit will diffuse until they are no longer identifiable as members of the comet's remains. But until that happens, we see them bunched into a meteor shower whenever they collide with the Earth.

The Earth passes through or very close to the orbit of Halley's comet every year, but the comet itself is in our vicinity only once every 76 years. Sometime in the next million years or less, however, this fiery spectacle will be stripped of all its ice and leave behind only a meteor shower. Many of these kinds of showers grace our skies now. One of the best known is the Perseid meteor shower, which we encounter every year about the eleventh and twelfth of August. It is not the richest shower, but since it falls during the height of the summer vacation season, when many are away from the lights and haze of cities, it is the most frequently observed.

Another noted recently is the Leonid shower, which passes us in mid-November. Unlike the Perseids, which are spread all along their comet's original orbit, the Leonids appear as a great shower every 33 years. This is because they are the remains of a comet with a period of that length and are still grouped where the comet would be if it were still intact. Were there a comet whose close approaches to the Sun caused it to spew its contents out into the interplanetary void, we might have an annual exhibit of celestial fireworks in the form of a tremendous meteor shower appearing for a day or two each year at the same time. In that event the shower would not be seen as a one-time event, but as a kind of dependable fireworks display, bringing about a festive air among many. If it should happen to coincide with the Fourth of July, Americans might consider themselves the chosen people under God even more than they do now.

These puny shooting stars are not like their much bigger relatives that appear singly and are of such size that the largest are not consumed within the atmosphere but can survive to strike the ground. Once on the ground they are known as meteorites, but while they are still aloft in space they are called meteoroids. The nomenclature divides those still whirling in space from those at rest on our planet.

Meteoroids proceed in orbits characteristic of asteroids, not comets. They usually strike singly or in groups of two or three, and the damage done is in direct proportion to their size. Fortunately, one large enough to create major damage is rare. Among the best known examples is the large meteoroid, or very small asteroid, as it could also properly be labeled, that struck northeastern Arizona some 50,000 years ago, leaving a crater about 4,000 feet wide and 600 feet deep, one of the youngest and best preserved meteor craters in the world and the most famous. The object would have destroyed a large city if it had landed there. Its diameter before impact was about 150 feet and it weighed a million tons or more, but even so it shattered on impact such that no piece of it remains bigger than about one ton. By now the crater has been mined extensively and the pieces of the intruder form souvenirs of considerable monetary value around the world; I own a one-pound piece of it and the observatory which I directed has one weighing around 370 pounds.

Comets possess one other property that greatly increases the likelihood of their possible collision with the Earth. Unlike the planets, whose orbits are nearly coplanar with the ecliptic and each other (see Fig.10.1), comets lack this degree of orbital decorum; their orbital planes appear at all inclinations from the ecliptic to its polar axis. With few exceptions, asteroids mimic the planets in their reasonable re-

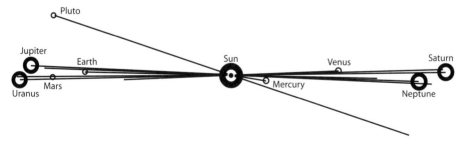

Fig. 10.1 The scheme of the solar system shown edge-on. Only Pluto moves at an orbit much inclined to the rest of the planets.

straint to inclinations close to the ecliptic. Since comets can approach from any quarter of the sky, and would also be detected as threatening to our continued well-being over a much shorter time, they may be more frequent intruders than asteroids of the same size. We must be on guard to a greater degree against cometic intruders approaching from every direction.

11

SEASONS OF
PARADOX

The Sun moves along the ecliptic by definition; it is the Sun's path traced out in the sky. Since the ecliptic is inclined $23\frac{1}{2}°$ from the celestial equator, over the year the Sun moves no closer than $90° - 23\frac{1}{2}° = 66\frac{1}{2}°$ from either celestial pole. It does so in June when it is nearest the north celestial pole and again in December when closest to the south celestial pole. The Moon and all of the planets except tiny Pluto move in orbits whose planes are inclined 7 degrees or less from the plane of the ecliptic (see Fig. 10.1).

What if one planet was not closely aligned with the ecliptic? We speculate here that one of our bright planetary neighbors happens to be in an orbit highly inclined to the rest of the solar system. This is extremely unlikely to have occurred in the process of formation. When the solar system condensed out of the nebula that gave it birth, it began to rotate. This is a result of the inherent turbulence of matter in unbounded space. Condensing as it did only caused the speed of its rotation to increase to the point where the centrifugal force balanced the condensation in the plane of the rotation, but perpendicular to it the collapse would have continued long afterward until it was condensed into a thin disk. It is not by chance that almost all of the mass of the system is nearly coplanar, for the Milky Way and other galaxies did much the same thing on a far grander scale. Disklike formations are much more common than globular ones, although some of the latter do exist. The objects that move in orbits highly inclined to the ecliptic all share one common trait; they are all very small and are all either comets or asteroids. They are the ones that could be tilted into a much different orbital plane by a near miss with a body of larger mass. This is why no massive planets are found well outside the primary plane of the solar system.

But what if somehow, from the gravitational influence of a passing star or something else, a bright planet did get dislodged from the plane of choice into an orbit that carries it near the poles, which one would it be? Not Mercury or Venus, because being closer to the Sun than we are, they wouldn't likely end up very far north or south of the ecliptic. That leaves Mars, Jupiter, or Saturn.

Let us suppose that Saturn does indeed move in an orbit inclined almost 90 degrees from our celestial equator. Saturn takes just about 30 years to circle the Sun once and shines between magnitudes −0.5 and +1.0, making it among the ten brightest objects in the sky most of the time. The planet is seen in the full phase since, like the other planets beyond the Earth, we can never get far enough from the Sun in angular terms to see more than a sliver of its dark side.

Saturn moves one-thirtieth of its way around each year, or about 12 degrees annually, and if it were in a highly inclined orbit, we would see it up near the north pole region for several years. Then it would proceed southward by nearly 12 degrees a year until it reached the vicinity of the south pole, where it would again linger. For several years around this time Saturn could not be seen at all from our midnortherly latitudes, but half a Saturnian year before and afterward, we would see it all night long every clear night near the Big and Little Dippers.

What would earlier societies have made of this anomaly? Would they develop some folk tale about the bad planet-god that was relegated or banished into its own kind of exile? Might Eudoxus just assign its crystalline spherical shell to its regular place between those of Jupiter and the fixed stars, even if it were upended along the way? Eudoxus, along with Aristotle and Plato, considered spherical shells, but later Hipparchus and Ptolemy required rings more or less in the plane of the system with their epicycles and deferents to carry the planets around the Earth. None of these geocentric models sheds light on the entire system until long after the time of Newton, when the dynamics of star formation were first recognized. Only then would this peculiarity be acknowledged for what must have occurred—something in the form of a mighty collision between Saturn and something large enough to whack this planet, 95 times as massive as our Earth, into a near right-angle turn.

The dynamics of the solar system require that almost all of its mass lie nearly in a plane. Only Pluto's motion is found in a plane angled more than 7 degrees from the ecliptic. In Pluto's case the inclination is about 17 degrees, still not very far from this basic plane. But when we consider the much smaller debris within the Sun's domain, we find a complete array of inclinations from 0 to near 90 degrees. As noted above, it is the small stuff in the system that gets knocked around by collisions or near misses with more massive chunks or even with each other.

When it comes to the alignment of the axis about which an object rotates, there is a general similarity between the planets but no firm rule. All objects rotate at some velocity; none happen to stand still in this regard. But the rotational speeds vary widely from about 10 hours for Jupiter to 244 days for Venus. Venus spins on its axis so slowly that it was almost the last planet for which we could determine the length of day.

Venus is a little smaller than the Earth. Although their surfaces and atmospheres differ almost as much as two nearby planets could, their interiors and seismic activity are similar. The Earth rotates such that a point on its equator covers its 25,000-mile circumference in 24 hours, or about 1,040 miles per hour, nearly twice the speed of a commercial airliner. Sluggish Venus takes 244 times as long to complete one turn on its axis, so its equatorial speed of rotation must be only around 4 miles per hour, the speed of a brisk walking pace. If one could exist on Venus and keep up a steady, quick stride along its equator with no breaks for window shopping, one would cancel out the rotation and stay in one place with respect to the direction to the Sun in the sky. Perpetual sunrise or sunset could be retained in this way. If the Earth were as leisurely in its rotation as is Venus, we can well imagine that life here would remain at the primeval level. Months of searing sunshine followed by the equivalent of a nuclear winter would render our planet as blighted as it was just after the mass extinction that finished off the dinosaurs 65 million years ago.

The direction of orientation of the axis about which a body rotates also varies considerably from one to another. Most adhere to one of two groups. The one group, of which the Earth is representative, has axial inclinations or tilts within about 5 degrees of our 23½ degrees to the ecliptic. Thus Mars, Saturn, and Neptune all have tilts of this general size, whereas the Sun, Mercury, Venus, and Jupiter barely have inclinations at all, 3 degrees or less. If our world had a tilt this small, it would appear upright, and globe manufacturers would not make their products hang over the way they do now. We would have no seasons to speak of, just a nice springtime condition all year long—rather nice as long as agricultural crops grow to harvest as they do. If, on the other hand, the orbit had a large eccentricity, the distance from the Sun would vary significantly enough for seasons to be reintroduced—with one impressive difference. On the Earth, the northern and southern hemispheres are out of phase; when it is winter in the north, it is summer in the south, and vice versa. Were the eccentricity to be the major cause of the seasons, the two hemispheres would experience summer together when the planet is closest to the Sun and winter when farthest. Jupiter is a case in point—with a tilt of 3 degrees and an orbital eccentricity of about 0.05, the latter would dominate the seasonal regimes on this giant planet.

That leaves Uranus and Pluto. Pluto's inclination is about 57 degrees, more than twice ours, and Uranus is the champion in this oddball contest with a tilt of 81 degrees, almost a full right angle. If our Earth were oriented as is Uranus, globes would be made to hang almost to the side, and the Sun's annual march would carry it to within 9 degrees of the north pole in June, after which it would plunge out of sight altogether from our midlatitudes sometime in October, not to return until February. Just as the Arctic and Antarctic regions do now, almost all the world would experience a midnight sun followed by months of darkness. Our seasons would be extreme; with four months of unending daylight, we would be cooked alive, only to exhaust our heating fuel and freeze during the coming winter. No thanks, these would not be good conditions for life as we have it here and now. It ranks down with asteroid collisions and Venus-slow rotation as hopeless for the higher life forms to evolve. More specifically, the tropical climatic regimes would experience their hottest weather in the spring and autumn, when the Sun would be near the celestial equator and thus nearly overhead at noon. At that time a city like Miami at latitude 26° north would exhibit a climate not unlike that of today, with the Sun up for half a day but rising to near the zenith, when its heat would be maximized. In June, when the Sun would be near the pole, it would shine for 24 hours each day but at a low elevation, and its heat would not be as excessive. In December, Miami would experience 24 hours of darkness, and the pervasive cold could well bring snow and wintry conditions even to southern Florida.

In New York, latitude 41° north, or London, latitude 51° north, the midsummer Sun would provide a more intense heat than in Miami because the Sun would be higher in the sky. Then in spring and autumn, the Sun would act about as it does now. The long winter would set in about October with the Sun low in the southern sky and come to an end in April, when it would advance rapidly from the southern latitudes. Winter would be more intense because the Sun would set in early November and not be seen again until late February. With darkness lasting for days, these northern cities would experience winters unlike anything since the last ice age, with numbing cold and temperatures approaching 100° below zero.

How did Uranus end up with such a crazy tilt? The most probable explanation is the catchall used for any anomaly found in our planetary system. We just surmise that some large mass collided with proto-Uranus in the early days of its formation. We know the celestial mechanics of any planet's orbit well enough to extrapolate backwards and forwards for thousands of years at least, but by no stretch can we do this for all 4.6 billion years or anywhere near it. So we can always speculate on peculiarities of orbit and orientation in the way of celestial billiards of one sort or another, and who is to gainsay our conclusions. Whether attempting to explain the tilt of Uranus or an imagined high orbital inclination of a large planet or even the birth of the Moon, we can postulate, "Oh, something must have just collided with it," as this is not only a plausible explanation, it is also probably true.

One more speculation has been made along this line. Most spinning in this system is done in one direction only; looking down from the north as we customarily do in portraying the planetary system, planets rotate and revolve in the counterclockwise direction. But a small minority proceeds clockwise or backward with retrograde motion from east to west. These are few in number; all planets revolve about the Sun in the normal direct or prograde way, but Venus, Uranus, Neptune, and Pluto rotate on their axes in the other, retrograde, direction, as do a few satellites in their orbits around their central planets. As the solar system formed, it—almost all of it—spun in the preferred direction but a few ended up going the wrong way on an otherwise one-way street. Something must have struck them early on and they reversed direction. The most interesting such case is Neptune. It spins backward and its one large satellite, Triton, also revolves against the grain.

One theory to explain this backed-up system is that Pluto was once also a large satellite of Neptune. This seems farfetched, as now the two cannot possibly collide or get entangled with each other—they are just too far apart, even though Pluto's eccentric orbit carries it closer to the Sun than Neptune for a few years in each of its long periods. Orbits are three-dimensional, and Pluto's orbital inclination keeps it well away from Neptune—now. But four billion years ago conditions might have been very different. We know that if both Pluto and Triton were in close orbits about Neptune, one of the two would likely have been ejected altogether. Three is a crowd in celestial mechanics if all three are of substantial mass, as these three were, and one will leave forthwith. In the course of Pluto's departure, the others might have gotten flipped around into the minority retrograde condition.

This theory was first published in the 1930s and caused such a furor that some astronomers were not thereafter speaking to each other, a situation that occurs among them now and then. The matter is still not and may never be settled to everyone's satisfaction.

12

MORE THAN ONE
PLUTO

Is Pluto a planet or is it something less? For years this question has been raised, discussed, and arbitrated again and again.

Pluto's 1930 discovery stemmed from the observations that neither Uranus nor Neptune followed the orbits that Newton's cosmos, even with Einstein's alterations, derived for them. Uranus strayed from the orbit it should have complied with, and this led directly to Neptune's discovery. But later observations confirmed that neither followed their orbits, hence another planet must be out there somewhere. This led Percival Lowell to his search, which succeeded only after he died in 1915. The discovery in 1930 by Clyde Tombaugh, a young staff observer at the time, of a starlike object, too small and too far away to be seen as a disk, was thought to satisfy the need for the restoration of Newtonian-Einsteinian order in the system. When Gerard Kuiper scanned the planet with the new 200-inch telescope at Mount Palomar soon after its completion in 1948, he measured its diameter at about 3,600 miles, placing it between Mars and Mercury in size, a respectable planet. Still, for years afterward, the three outer planets could not make up their minds to behave.

For an outer planet to make the others follow the gravitational rules as computed at the time, an object of about the mass of the Earth was required. In order for Pluto to do the job, it had to be either by far the densest or the blackest body in the universe; the densest in order to pack an Earthlike mass into a body smaller than Mars, or the blackest with an absurdly low albedo that could make it appear so faint. But Pluto turned out to be neither the one nor the other. Thus another, tenth, planet, Planet X, must be still farther out there to handle the gravitation thing. The situation remained in a muddle as no new planet could be found until the second *Voyager* space probe passed near both Uranus and Neptune by 1989, providing the first clear images of them and their moons and redetermining their orbits with far greater precision than earthbound observations could muster. Astronomers at the Jet Propulsion Laboratory in Pasadena used new observations made by the flyby and came to the conclusion that these two giants were indeed following the orbits we designed for them, and as a result the need for Planet X vanished! The discovery in 1978 of Pluto's satellite, Charon, gave us Pluto's mass with the familiar story applied from Kepler's harmonic law with masses. A puny mass, one-sixth that of the Moon, and a diameter of somewhere around 1,400 miles raised the question: was this a proper planet or something less—an asteroid? The plot thickened when other Plutos were found out there, none quite so large but of the same constituents and in very similar orbits.

In 2001 and again in 2002, we have found two more Plutos, or near enough. They have been known by their designations in astronomical parlance, 2001 KX76 and 2002 LM60, and as soon as their orbits have been calculated with a precision that enables us to recover them as we choose, they shall receive names. They are tentatively called Ixion and Quaoar, respectively, after gods, since names for these objects generally favor the deities of any of a number of cultures. Ixion's size is uncertain, but it is probably larger than Charon, at 600 miles in diameter; Quaoar is certainly larger, being about 800 miles across, making it more than half the size of Pluto (1,400 miles) and the largest object discovered in the solar system since Pluto itself was found in 1930. Quaoar has a nearly circular orbit with a period of 288 years and an orbital inclination of about 8 degrees. Compare with Pluto's 248 years and 17-degree inclination.

As John Davies has written, the past decade has seen a doubling in the size of the solar system. The area beyond Neptune, now called the Kuiper belt, after Professor Gerard P. Kuiper, who was among the first to call attention to it, is home to a group of minor planets discovered since 1992. This recent development has dramatically broadened and altered our understanding of how the solar system was formed and has provided answers as to the origin and nature of Pluto. Theoretical physicists had decided that there must be a population of unknown bodies beyond Neptune, and a small band of astronomers with access to very large telescopes set out to find them. What they discovered was an extended ring of minor planets whose orbits and physical properties are far more complicated than anyone expected. Pluto is simply the largest known of these Kuiper belt objects, or KBOs, playing a role analogous to that of Ceres, the largest object in the closer asteroid belt. It is entirely possible that a few KBOs larger than Pluto still await discovery.

Is Pluto then a planet or is it not? If its very small size were known right after its discovery, or if we had known that it cannot account for discrepancies in the orbits of the next two outermost planets, then probably not. It is a trans-Neptunian object or a KBO, the first discovered of a new type of animal, with its moon, Charon, being the second. The next two in size, Ixion and Quaoar, are also larger than either Charon or Ceres, and another, Varuna, is about as large. If these three with Pluto and Charon, the five largest KBOs, were located in the closer asteroid belt along with Ceres, what could we see? These are icy worlds, and some have higher albedos than Ceres and most of its rocky neighbors. Pluto, the brightest, were it in Ceres' orbit, would appear to us at the third or the fourth magnitude at opposition, easily visible to the naked eye and certainly known to ancient civilizations as a planet. But it would not be a dirty snowball if it were that close—the icy portions of Pluto and the other trans-Neptunians would have sublimated, and in their places might be meteor showers. Occasionally, one of the smaller of these objects is perturbed in such a way as to fall into the inner solar system. When that happens we see a comet. Halley's comet, now in a very eccentric and perturbed path, could once have been a KBO. In Pluto's case, the terminology of its class of object is more a question of semantics and habit than of astrophysics.

Pluto and the largest inhabitants of the Kuiper belt could not exist in the inner solar system; they certainly could not maintain their integrity. If, then, Pluto were to find itself in the midst of the main asteroid belt, the one between Mars and Jupiter, we would behold not a fairly faint star, but just possibly a world on its way toward dissolution. At this distance comets begin to grow tails due to the rising temperature enforced by solar radiation; the Sun's energy causes the icy majority mass of the comet to sublimate, and particles of gas and dust are emitted and blown away by the out-rushing solar gas known as the solar wind. The solar wind always forces the particles generally in the direction away from the Sun, and these form the tail or tails (for comets can have more than one tail); regardless of the comet's own orbital motion, the tail streams away from the Sun. The gaseous material forms a tail that is blue in color, whereas its dust, the stony minority, appears as a yellowish pink tail that can sometimes be displaced from the tail of gas. The largest comets on record originate as roundish objects, perhaps 20 miles in diameter. Comet Hale-Bopp of 1997 was such a comet. Its large size allowed it to be spectacular even at its closest distance of nearly 1 AU, where it was seen in the evening sky by tens or even hundreds of millions of people. Whether much larger objects would act as jumbo-sized comets is difficult to say. Nevertheless, sublimation would affect even the largest icy objects as they approached the Sun, and some emission would be bound to occur.

Note. At this writing (in March 2004), in a paper entitled "Discovery of a candidate inner Oort cloud planetoid," submitted to *Astrophysical Journal Letters* by M. E. Brown, C. Trujillo, and D. Rabinowitz, they report the discovery of an object that may increase the size of the solar system by an order of magnitude and confirm the existence of the Oort cloud. This object has been assigned the name Sedna, after the Inuit goddess of the sea, and, at about 1,000 miles in diameter, appears to replace Quaoar as the largest object discovered in the system since Pluto. Its present distance is given as 90 astronomical units, not far beyond its perihelion distance near 76 AU, about twice the orbits of Pluto and Quaoar, but its period has been estimated at a whopping 10,500 years!

We can again use the harmonic law to quickly estimate its average distance from the Sun to be 480 AU in a highly eccentric orbit (about 0.84 for the eccentricity) with an aphelion distance of almost 900 AU. This amounts to about 0.014 light years or 1/300 of the way to Alpha Centauri. Sedna is the first-discovered member of the Oort cloud.

As the discoverers of Sedna point out, the planetary region of the solar system, defined as the region including the nearly circular low-inclination orbits, appears to extend to the outer edge of the Kuiper belt at about 50 AU from the Sun, as mentioned in Chapter 10. Many bodies in highly eccentric orbits from the planetary region—comets and scattered Kuiper belt objects—cross this boundary but all have perihelia *within* the planetary region. Far beyond this edge lies the realm of the comets, at around 10,000 AU and beyond, in the Oort cloud. Although most stay in the Oort cloud indefinitely, it is in this distant region that gravitational perturbations by nearby passing stars and tides due to the massive center of the Milky Way galaxy cause them to enter the inner solar system, where they are discovered. "Every known and expected object in the solar system has either a perihelion in the planetary region, an aphelion in the Oort cloud region, or both. . . . The orbit of this object is unlike any other in the solar system. It resembles a scattered Kuiper belt object but with a perihelion much higher than can be explained by scattering by any known planet. The only mechanism for placing the object into this orbit requires either perturbations by planets yet to be seen in the solar system or forces beyond the solar system."

In Chapter 4 we found Proxima Centauri to be about 13,000 AU from the brighter pair, Alpha Centauri A and B, still well beyond Sedna's maximum distance from the Sun but consistent with the Oort cloud. If such a star were truly part of our system, as we supposed, most of the comets in the cloud would have long been scattered in all directions.

III
THE STARS

13
WHAT IF THE SUN WERE RED?
OR BLUE?

Stars come in an assortment of colors. Not just any old color, as crayons do, but in the colors of the spectrum or rainbow. In Chapter 2, we noted that as stars get hotter they proceed in tint from red through orange, yellow, and white to blue. There are instances in the past in which observers have described the colors of some individual stars as mauve, lilac, garnet, and heliotrope, but no one today assigns to stars colors of such fancy; these tones are left for interior decorators. But the two extremes, red and blue, are very real, as anyone familiar with the constellation of Orion can attest. That brightest of all constellations has two zero-magnitude stars, two of the ten brightest in the sky. Both are supergiants, among the most brilliant stars in our galaxy. Betelgeuse in the upper left shoulder is about as red as stars get, and Rigel, representing the knee to the lower right, is distinctly blue. If Rigel is not bright enough for you, try Sirius, just south of Orion and to the left. That brightest of all stars is the most luminous blue object in the heavens, and Betelgeuse is the brightest of the red objects except for Mars during one of its favorable orientations. What makes stars shine in distinctly different colors? Astrophysicists found that color is a function of surface temperature; the hotter the star, the bluer its color.

The Sun is a proper middle-of-the-road yellow star, neither too hot nor too cool, but, as Goldilocks found the possessions of the baby bear, just right. It is best that this is so, for a blue star shines too much in the lethal ultraviolet region, and a red star gives off too much in the red and infrared regions. As it is, our star emits quite enough ultraviolet radiation, which can lead to melanoma and other skin problems through overexposure. Without the ozone layer in our upper atmosphere, we would quickly feel the very adverse effects of a solar-induced overdose.

When we speak of light, it can be with either of two meanings. Light as our eyes see it is limited to the colors of the rainbow: red, orange, yellow, green, blue, and violet or purple. But our eyes are sensitive to only a narrow range of color or, more properly, wavelength within the great totality of wavelengths. The difference between the colors lies only in the length of their waves, with violet being the shortest and red the longest (refer back to Fig. 2.4). The difference in length is less than two to one, but on both sides light of much shorter and much longer wavelengths exists in abundance. Past (that is, shorter in wavelength than) the ultraviolet region lie x-rays and gamma rays, and beyond the infrared region we have radio waves, including the AM, FM, and short-wave radio bands. Taken together, the totality of wavelengths is referred to as the electromagnetic spectrum. To almost all these colors our eyesight is insensitive, but where it counts, where the Sun shines, we can see full well. More than 92 percent of the total energy radiated from the Sun falls within the visual portion of the entire spectrum. The evolution of life has been oriented to the maximum intensity of sunlight, as we would expect.

If we once again consider the Hertzsprung-Russell or HR diagram (see Chapter 2), in which color or surface temperature is plotted against total luminosity of a star, we see that the main sequence, defined by the great majority of stars, is a sequence not only in mass, but also in age or expected lifetime. For example, the Sun is close to 5 billion years old. Astrophysicists know that it will live for about another 5 billion years. Then what happens? The mighty source of solar energy is known to derive from the thermonuclear fusion process, featuring the conversion of hydrogen into helium. This goes on in the solar core, where temperatures are at their hottest. Four atoms of hydrogen meet to form one atom of helium with a little mass left over. This mass gets converted into energy, and some of it passes from the deep core of the Sun, where it is hot enough for this to happen, out into space.

But as Einstein showed with his famous equation $E = mc^2$, a tiny bit of mass converts into a whole lot of energy; thus, a hydrogen atom weighs 1.008 on the atomic weight scale, and four of them weigh 4.032. One atom of helium weighs 4.003; thus each transformation, called fusion, loses the difference, 0.029, as energy. This same thermonuclear process is what fuels a hydrogen bomb, and, like the center of the Sun, the fusion can only take place at a temperature of around 15 million degrees Celsius, actually Kelvin, or more. The trigger for the heat necessary to detonate the H-bomb can be a fission bomb of uranium or plutonium.

Now a star twice as hot in surface temperature as the Sun appears much brighter and blue in color. Two such hotter main-sequence stars

TABLE 3. MASSES AND LIFETIMES OF REPRESENTATIVE STARS

Mass	Example	Total Lifetime as a Star (Years)
15	Spica	10 million
6	Achernar	100 million
2	Fomalhaut	1 billion
1.0	Sun	10 billion
0.4	61 Cygni B	100 billion
0.15	Barnard's Star	100 billion +
<0.08	Brown Dwarfs	About one billion
0.001	Jupiter	Will not shine at any time

are Sirius and Vega. Each is about 2 to 3 times as massive as the Sun, but Sirius is about 25 and Vega 50 times as luminous, pouring out 25 and 50 times the energy output of the Sun in the same interval of time. It is not hard to see that they will run out of energy well before the Sun does; in fact their total lifetimes are only some half billion years, not 10 billion like the Sun. The lifetimes of several stars, as well as their mass in terms of that of the Sun, are summarized in Table 3.

Brighter blue stars, such as Beta Centauri and Spica, will live as stars for only some 10 million years, and Fomalhaut, a star a little fainter than Sirius and Vega, for a billion years. The Sun we know has a total lifetime of 10 billion years, and the faint red dwarf stars 61 Cygni A and B and Barnard's Star will last for 100 billion years or more, but that is to come. No star is older than about 13.5 billion years; it was that long ago when the first stars formed soon after the Big Bang. As we have seen, stars smaller than 0.08 solar masses, or 80 times the mass of Jupiter, cannot reach core temperatures hot enough for fusion to begin, and after a comparatively short lifetime as a star, through gravitational contraction, they fade away into darkness. It is in the region of stellar mass between 0.08 and about 0.4 or even 0.5 times the mass of the Sun that all stars now shining will continue to do so for much longer than the time elapsed since the Big Bang.

What about a little red dwarf star such as Proxima Centauri? This star has maybe one-tenth of a solar mass (the Sun by definition weighs one solar mass) but shines with less than one ten-thousandth of the Sun's energy. Clearly, Proxima is good for tens if not hundreds of billions of years to come. We believe the universe to be about 13 billion years old, 13.7 billion years since the Big Bang, according to new information, making it just three times as old as is our solar system. All stars like Proxima that formed anywhere in any galaxy are still alive, shining along in their minuscule fashion. Brighter stars like our Sun that formed during the first few billion years after the Big Bang have evolved, as stars of this mass will do, into a brief red giant status, and then lose their envelopes or outer regions while their cores condense into white dwarfs or degenerate stars. These are stars in which the atomic nuclei are all squeezed together such that a cubic inch of this degenerate matter would weigh many tons. Since stars like Vega and Sirius last less than one billion years, most stars of their kind have already made it into the stellar graveyard.

A typical white dwarf is Sirius B, the faint companion star to Sirius. It is 1/10,000 as bright as Sirius and appears to us at magnitude 8.5, even though it has nearly the same surface temperature as does its brighter companion. A star of this brightness would be visible in binoculars were it not for the fact that Sirius is so close and so bright that resolution is very difficult. It is so difficult that its existence was first detected by Bessel in the 1840s from the gravitational perturbations it imposed on Sirius, which Bessel noted wobbled with a 50-year period in addition to its parallactic and proper motions. He accepted the only realistic explanation: namely, that a companion of low luminosity but with a substantial mass swung in a 50-year orbit with Sirius. Later, in 1861, Alvan Clark, the renowned American lens maker, detected the companion while testing a new lens for a 16-inch refractor. This degenerate object is blue like its brighter companion but it is far too faint to be a main-sequence star. It is about as massive as the Sun but about the size of the Earth. The only explanation for the high density it must have is that it consists of atomic nuclei jammed against each other with no room for electrons to circle about them.

Now we see the nature of the problem encountered when we postulate a civilization like ours on a planet orbiting Vega. Life on the Earth formed not long after the condensation of the solar system out of its primordial nebula, but all the higher life forms, animal or vegetable, didn't come along until about half a billion years ago, probably before Vega's formation. Would life ever have time to evolve and develop if its star only existed for that length of time? We can't say with certainty, but the highly probable answer is no!

Stars like Proxima do live long enough to see any kind of evolution we can imagine, but here we encounter a different problem. Defining life as life-as-we-know-it, as we can only assume, we find that it needs temperatures rather like those encountered here on Earth. Mercury and

Venus are far too hot for carbon-based molecular structures to hold together, whereas at the asteroids and beyond it is much too cold. Only the Moon and Mars have any chance of supporting creatures like us, but being small and airless or almost so, they could neither form nor sustain us.

We end up with an empty spherical shell, hanging in space and centered on the Sun, whose closest surface lies between Venus and the Earth and whose outer surface is near Mars, within which life could exist, a kind of habitable zone. But the heat from Proxima is so small that its habitable zone lies almost on its surface, where no planet could maintain a separate existence. Space beyond this miserably tiny habitable zone would be far too cold for life. It may seem a tautology to state that the stars most likely to support life are stars rather like the Sun. But the physics of the problem just doesn't seem to allow any other conclusion.

The main sequence of stars on the HR diagram is indicative of one more stellar property in addition to a sequence in mass, luminosity, and longevity. It is known to be a sequence in the frequency of each kind of star. We can see this as an enormous selection effect just by considering the twenty stars of the first magnitude or brighter. Every one of these twenty is intrinsically brighter than the Sun, and the majority is more than 100 times as luminous. In fact, all stars of the third magnitude or brighter, nearly 300 stars altogether in the whole sky, are more luminous than the Sun. Considered by itself, Alpha Centauri B shines at the first magnitude but is intrinsically fainter than the Sun; its light combined with the brighter A makes the pair, seen as one star by the naked eye, intrinsically brighter. Yet after a great amount of work searching for even the nearest of faint red dwarfs like Proxima, we are left with the inescapable fact that about 90 percent of all stars in our galaxy or elsewhere are fainter than the Sun! Thus the bright stars are as lighthouses among fireflies, great beacons that can be seen for hundreds of light years in every direction. This while Proxima at four light years cannot be seen from the Earth even with binoculars.

The situation has been likened to the tremendous difference between a sample of, say, the one hundred people whose names appear most frequently on the front page of a newspaper (presidents, senators, sports heroes, film stars, and the like) and the hundred who live closest to you in your neighborhood. Even if you live on Fifth Avenue or in Beverly Hills, it is highly unlikely that even one name would be found on both lists. So it is with stars. Our premise that we might have been formed on the planet of a very red star or a very blue one is flawed, probably beyond repair. Stars like Alpha Centauri A or B, a little brighter and fainter than the Sun, might work, but none far from us in luminosity in either direction is at all likely to fill the bill.

We are not yet able to detect planets as small as the Earth around any other star, but we know of more than one hundred of the size of Saturn or Jupiter or even larger outside our solar system. Of those, all or almost all accompany stars not very different from the Sun. We haven't a large enough sample yet to make more than a very preliminary estimate, but it appears that something like 7 percent of solar-type stars have at least one planet.

If the Sun were moderately redder or bluer than it is, our eyes would have adapted to account for it. As it is, the Sun emits maximum light and energy at about 550 nanometers, or nm, where one nm equals one-billionth of a meter. This is the length of a single wave of light from one crest to the next in the yellow portion of the spectrum. Our eyes can detect wavelengths from about 400 nm at the edge of the violet end to nearly 700 nm at the red end of the visible spectrum. The stars likely to be central stars with higher life forms shine at their maximum well within this range from about the blue-green at 450–500 nm to the orange-red near 600 nm. We and other animals would see a little more of the ultraviolet in the one case and a bit into the infrared in the other. Since the impression of color is one of those things that one cannot describe to another, we can only guess what colors we would see and how we would see them.

14

THE VERNAL EQUINOX
LIES IN VIRGO

The sky changes slowly. With an occasional exception, alterations among the stars and constellations are orderly and require centuries for major changes to be detected. The supernovae that appeared in 1572 and 1604, known as Tycho's Nova and Kepler's Nova, were and would be today totally unanticipated. But the other changes, the motions of the Sun and planets, eclipses, transits, and occultations are now predicted well in advance, if only because they are all within the solar system. Predictions that involve objects belonging to the solar system are based on the high level of celestial mechanics that is possible when masses and distances are so well known. Outside it, the data are just not good enough to make for the levels of exactness we associate with the planets.

Among the appearances of the stars, we have mentioned two long-term alterations. One reorders the star patterns we call constellations, and the other leaves them fixed but changes their orientation with respect to the observer and his horizon. This latter is the motion we call precession of the equinoxes, or more commonly just precession.

Precession, as explained earlier, is a motion of the Earth, and thus slowly and steadily it reorients the sky as a unit with all of its stars moving in lockstep. It is caused by the fact that the rotation of the Earth gives it an equatorial bulge and makes it slightly oblate or flattened at the poles. This bulge is pulled on by the Sun and the Moon and very minimally by the other planets. The Earth rotates like a gyroscope that also has extra mass at the rim of its equator (see Fig. 14.1). External gravitational force wants to pull the Earth or the gyroscope over onto its side, but it turns out that the globe and the gyroscope end up spinning with a slow change in the direction of the axis of rotation.

Hipparchus discovered the precession in the second century B.C. from a comparison of the positions of stars in his catalogue with those

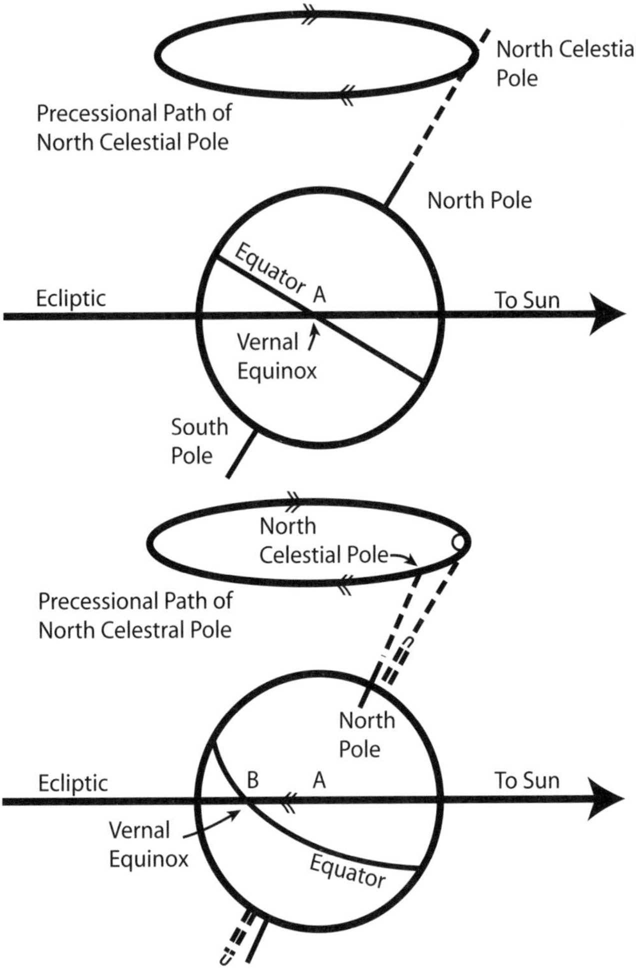

Fig. 14.1 Precession in the motion of the Earth.

of an earlier astronomer, whose catalogue is now lost. He noticed a uniform shift to the east over the intervening centuries, but it was left for Isaac Newton to derive and explain this movement some eighteen centuries later. Newton showed that the precession confirmed the existence of the bulge at the equator before it was known or measured, but once he predicted it, it was sought and detected. This bulge is only 0.3 percent of the diameter (7,927 miles through the equator and 7,900 through the poles). The action upon it by the gravitation of the Moon and the Sun pulls it sideways like a top.

The period of the precession, the time it takes for our axis to return to a point, is 25,800 years. This means that Polaris, a bright second-magnitude star (at magnitude 2.0, about the fiftieth most luminous star in the sky) that today happens to be located near the celestial north pole, around which the sky appears to rotate, will drift well away from its favored status as our north star or pole star, but in 25,800 years it will return to this spot in the sky. During the intervening eons other stars will at times pass close enough to the pole to be thought of as pole stars. After Polaris, the best known of these is Thuban, a star of magnitude 3.6 in Draco, the dragon, which was near the pole almost 5,000 years ago when the great pyramids were being built (see Fig. 14.2). About as many years from now, near A.D. 7000, Alderamin, the brightest star in Cepheus, the king and husband of the more luminous Cassiopeia, will have its turn. And about 13,000 years from now, which means also 13,000 years ago, when the last ice age was near its end, the bright star Vega did and will become the brightest among the north

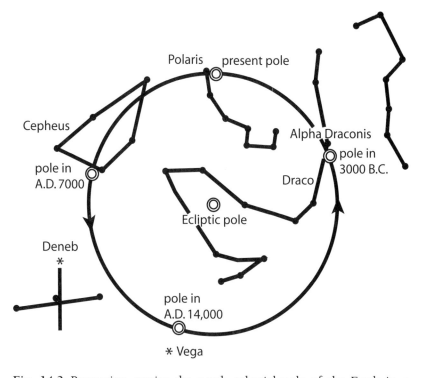

Fig. 14.2 Precession carries the north celestial pole of the Earth in a circle in 25,800 years. The pole was near the star Alpha Draconis (Thuban) 5,000 years ago and will point to Alpha Cephei 5,000 years from now. It is now near Polaris, and 12,000 years hence, Vega will be the pole star.

stars we can have. At the moment no significant star marks the south celestial pole, but in times past and in times to come, one or another bright star will pass close to it.

The amplitude in the declination of a star (the angle the star makes with the celestial equator, analogous to latitude on the Earth) changes by twice the amount of the inclination of the Earth's axis, or 47 degrees. Stars, except for a few near the poles, go through this swing in the 25,800-year period of the precession.

Let us look at the two constellations favored by the largest numbers of very bright stars, Orion and Scorpius. Scorpius has four of the fifty stars in the sky that are brighter than magnitude 2.0 and Orion has five. Both are composed mostly of highly luminous supergiant stars on the order of a thousand light years or more from us, because each is essentially a star factory in which stars are still being formed out of the interstellar nebular material lying in abundance in these regions. In a later chapter we shall see other similarities between these two luminous star groups. As we saw earlier, supergiants are very short-lived stars, as they convert their hydrogen into helium and energy so quickly. One of the brightest stars in each constellation is red; Betelgeuse in Orion and Antares in Scorpius are the two brightest red stars in the sky.

We see brilliant Orion in the wintertime about halfway up from the southern horizon to the zenith, and by springtime it is gone into the twilight glare. Scorpius is almost exactly opposite Orion in the sky, and as Greek mythology makes clear, they are that way because Orion, the mighty hunter, was stung to death by the scorpion, and the gods placed these enemies in the sky as far removed from each other as possible, each rising as the other sets. That makes Scorpius a summer constellation seen at its best low in the south from our midlatitudes in June and July. Scorpius lies so far south that we never see it in its full splendor from the latitude of New York, but in 13,000 years, half a precessional period ago or from now, it will be the scorpion that rides high in the winter sky and Orion will appear in the southern summer sky, so low that from New York only Betelgeuse and one other bright star would be visible.

Precession enters into astrology in a subtle way. Astrology, Western astrology, divides the 360 degrees of the ecliptic into twelve equal segments called signs, each 30 degrees in length. The area of sky extending 8 degrees on either side, north and south, of the ecliptic is defined as the zodiac. The zodiac, so named because most of the signs represent animals, signifies the region in which all planets but Pluto must appear all of the time. Among the planets only Pluto, discovered in 1930, too late to redefine this zone, can drift more than 8 degrees from the ecliptic at any time.

The twelve signs that form the background for the ecliptic bear the same names as do twelve of the constellations in the same general area of the celestial sphere. In the traditional order, eastward from the location of the vernal equinox, they are as follows:

> Aries, the ram
> Taurus, the bull
> Gemini, the twins
> Cancer, the crab
> Leo, the lion
> Virgo, the virgin
> Libra, the balance
> Scorpius, the scorpion
> Sagittarius, the archer
> Capricorn, the sea goat
> Aquarius, the water carrier
> Pisces, the two fish

In classical times when most astronomers doubled as astrologers, each of the twelve constellations lined up closely with the sign of the same name. In the twentieth century astronomers have made areas for each of a total of eighty-eight constellations in the heavens so that every point is in one or another of them, but in those days the constellations consisted of star figures grouped by stars that in someone's imagination appeared to represent a person or an animal or occasionally an inanimate object, but only now and then was the resemblance at all close. There was no overall agreement between people or societies as to just which stars belonged to a particular star group; thus the sizes of the groups were indistinct. A consensus formed as to the location of the brighter stars, those of about third or fourth magnitude or brighter, but fainter ones were sometimes placed within one constellation and sometimes in another. As a result the signs in classical times fit the constellations more or less, but no one allowed for the effect of precession upon the locations of either set. It was Ptolemy, more than any other ancient scholar, who in his book, *Tetrabiblios*, laid down the framework for astrology as it is practiced even today. The problem introduced by precession is that in every 25,800 divided by twelve, or 2,150 years, the constellations have shifted by one 30-degree segment but the signs have not.

Now almost 2,150 years later, these star groups are out of phase with their respective signs by about the length of one sign. The signs line up such that Aries, the ram, is the sign in which the Sun passes the celestial equator heading north each March 21; thus Aries contained this point known as the vernal equinox in Greco-Roman times. Similarly the three other principal points, the autumnal equinox passed by the Sun every September 23, and the two solstices, of the summer (June 21) and the winter (December 22) fell in the signs of Libra, Cancer, and Capricorn, respectively. But today each of the four points and their respective signs have backed up into the constellations immediately to the west of the former ones. The equinoxes have backed into Pisces and Virgo and the solstices into Gemini and Sagittarius. With few exceptions, the astrologers inaccurately retain the old signs. To be sure, both the signs and constellations are entirely arbitrary; neither has an origin that is connected to the sky. However, the fact remains that for its own internal consistency, astrology calls for each person to apply the horoscope of the sign preceding the one indicated in the list above. Thus, Leos are really Cancers, and Aquariuses are in truth Capricorns. From the proper point of view for these times, Henry Miller would have written books entitled *The Tropic of Gemini* and *The Tropic of Sagittarius*. Were there any external truth to this business, the matter might be of the greatest of human concerns. The precession will continue to act to move the solstices and equinoxes; the vernal equinox will soon back from Pisces into Aquarius, the event giving rise to the expression "the Age of Aquarius." And every 2,150 years or so it will move into another sign, eventually reaching Virgo in 13,000 more years, and back to Pisces 13,000 years beyond that.

The length of time for the vernal equinox to shift from one sign to the next being 2,150 years, this has become in some circles a significant period. Oswald Spengler, in his *Decline of the West*, found this period to be the interval between the lives of his equivalent persons in different civilizations; thus Napoleon and Alexander, comparable figures in the histories of their respective societies (in this case, the modern and the classical), are spaced by this interval of time. As Spengler portrays it, we can look for someone comparable to Julius Caesar to emerge about a century from now. This note of mysticism is one of the reasons he and other cyclic historians have lately fallen into disfavor. Others have added to the significance of the "great year," the 25,800-year precessional period and its dozen divisions, in ways that are for the most part pseudoscientific in nature. For one example, the Sphinx poised eternally near the great pyramids of Gizeh, near Cairo, Egypt, is portrayed by some as sharing the body and tail of Leo, the lion, with a head that represents Virgo. Half a great year, 13,000 years ago, the vernal equinox lay between these two adjacent constellations, as it does between Pisces and Aquarius today. The autumnal equinox lay then where the vernal equinox does now and vice versa—suggesting that the Sphinx was created much more than 4,500 years ago when the pyramids were built, perhaps by thousands of years. Little credence is given to this theory by archaeologists.

One of many such references along this theme is the book *Hamlet's Mill*, by Giorgio de Santillana and Hertha von Dechend. The basic premise of this extended essay is that ancient myths "can be interpreted as a code language expressing the astronomical knowledge of precession of the equinoxes among ancient cultures." The transmutation of the vernal equinox from one arbitrarily constrained sign to another serves as a guide to understanding primitive mythologies, including those of the ancient Hebrews and their monotheistic beliefs, which gave rise to one of the greatest motivating factors of the modern era. The admixture of the "great year," that very uniform precessional period, with the belief in a single deity, the presumed existence of vast, timeless aeonian civilizations that could grasp the essentials of the precessional motion (curiously, without noticing the uniquely irregular and unsettling proper motions of the stars), blended in some sources with the legend of Atlantis and a radically earlier creation of the Sphinx, has, if nothing else, promoted a cottage industry in the publication of a network of books in the same vein for the gullible among us.

Precession alters the position and rising time of the constellations, but it does not change their size and shape or their orientation with respect to each other. How long will our familiar constellations last? How many thousands of years must pass before all of our star groups are so distorted by the individual proper motions of their member stars that no one living today could recognize any of them? We can see (in Fig. 14.3) the Big Dipper, our favorite asterism, as it looks today and as it will appear 100,000 years in the future. It seems that the recognition of the Dipper has endured from before the end of the last ice age and well before recorded history, and is likely to last into the next ice age at least.

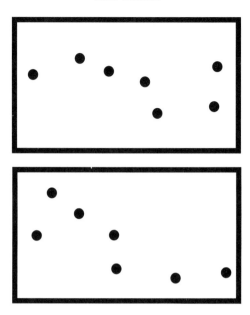

Fig. 14.3 The Big Dipper or Plough as it appears today and as it will appear in 100,000 years. Unlike precession, the proper motions of stars will come to distort the shape of a constellation.

In the year 1718, Edmond Halley became the first to find that stars moved, each with its own individual motion across the sky, called proper motion. After taking account of the precessional motion, he noticed that the bright nearby stars Sirius, Procyon, and Arcturus had shifted about a degree with respect to their stellar neighbors in the intervening two millennia since Hipparchus. In 25,800 years the shape of many star figures will be noticeably altered. Sirius and our other nearest stellar neighbors may be well away from their present locations by that time, and some of them are receding toward a greater distance from us and may appear as nondescript fainter stars. Others now inconspicuously approaching us may shine more brightly in their places, but a few asterisms will remain recognizable for much longer. Above all, two of our brightest, Orion and Scorpius, will retain their present shape for the better part of a million years.

As we will see in later chapters, if either we had no moon and the precessional period were many times as long as it is, or if we were farther removed from the Milky Way and had no other stars nearby, we would view the whole matter of long-term stellar motions very differently—even among those titillated by Atlantis and the "great year."

15

VEGA AND DENEB
CHANGE PLACES

The summer triangle consists of three bright stars in three different constellations; they dominate the zenith area in the sky all summer long, and with Arcturus to the west and Antares in the deep south, they are the brightest stars visible at the time. All three are blue, with Vega easily the brightest at magnitude 0.0, while Altair and Deneb, at 0.8 and 1.3, respectively, are not much fainter. In the days before excessive light pollution in the cities and suburbs, they were seen along with their constellations rather easily, and the term "summer triangle" was rarely used to denote them. Today, however, with the extensive light pollution affecting the night sky for most of us, these three bright stars are often the only conspicuous stars overhead in the summertime, and this term has come into widespread use. Vega is part of Lyra, the harp; Altair lies in Aquila, the eagle; and Deneb is in Cygnus, the swan. Although all are blue stars, they are very different, Vega and Altair being normal main-sequence stars and Deneb being altogether much more luminous. Altair is very close as bright stars go—it lies only 16 light years away, whereas the brighter Vega is 26 light years off. Since Vega is more distant and yet appears the brighter star, Vega is truly several times as luminous as its neighbor.

Deneb appears a bit fainter than either of the others, but in reality it is most assuredly not. Deneb is still too far for a reliable parallax measurement of it, but from its spectral and astrophysical properties, it must be some 1,500 to 3,000 light years from us and is therefore likely to be the most distant of all of our first-magnitude stars. We see it now as it appeared in ancient or early medieval times, but astrophysicists assure us that it probably appears the same as it would if we were able to see it as it exists today. To be bright in our sky from such a distance, Deneb must be a truly brilliant source of light and energy indeed, a veritable searchlight shining amidst a field of fireflies. This supergiant has to be one of the most luminous stars in our entire Milky Way galaxy, an assemblage of several hundred billion stars in a disk some 100,000 light years in diameter. Deneb puts forth more light intensity in a single night than does the Sun in an entire century! Here is a star with about 20 or 30 times the Sun's mass but emitting tens of thousands of times as much energy. So for a few million years it will put on a real show, but then in a twinkle it will briefly turn red, resembling Betelgeuse and Antares, and then—literally—blow up into a supernova, as Tycho's and Kepler's Novae did four centuries ago. A supernova is not like a nova, not at all. Novae are limited to certain types of luminous stars that shed an outer shell of material every so often, every few thousand years, for example. But a supernova blows the whole star into smithereens, filling the surrounding space with pieces of its innards; there are no repeat performances because there is no more star. Only very massive stars do this; stars like the Sun are much more properly behaved and do not go about blasting their interiors into their part of the universe. A supernova is an entirely different type of phenomenon; it is so bright that for a few days or weeks it can almost outshine the rest of its galaxy.

If Deneb were as close as Vega or Altair, it would appear about as bright as the Moon at first or last quarter, but would still be seen as a point of light, not a disk. At its present location in Cygnus it lies so far north that we would never have a completely dark night in the higher midlatitudes. One could always see one's hand in front of one's face, at least on nights that were not fully overcast. If Deneb were much closer and brighter, everyone from Little Red Riding Hood to Frankenstein's monster would be able to navigate at night, and our ghoulish literature might have to be modified. Count Dracula might also experience problems maintaining his bizarre circadian life style, jumping into and out of his coffin with dawn and dusk.

That's the good news. The bad news comes when a nearby supergiant star decides to become a supernova. Deneb is probably not yet about to do this, but Betelgeuse just might. Betelgeuse has already gotten into its red stage, and at some point it will suddenly flare up to a light intensity far beyond that of the full Moon, whereupon it will be brightly visible by day and by night, and we will experience a sky of twilit intensity after sunset. This might happen next Friday afternoon or not for another 10,000 years; we can't call it better than that. Less massive stars such as the Sun or Vega will throw moderate tantrums, called novae, in due course of time, but only a small portion of their mass will be thrown out into space. The supergiants, however, including Betelgeuse, Rigel, Canopus, Antares, and Deneb, will not go gently into that good night; no, they will blow up and splatter themselves all over their corners of our Milky Way galaxy. These most violent cataclysms in all of nature are not common; the last five occurring in our entire galaxy were first seen in the years 1006, 1054, 1181, 1572, and 1604, the last two being known as Tycho's Nova and Kepler's Nova (the distinctive nature of supernovae as differentiated from novae was not at all understood at that time). These five most recent supernova events in our galaxy were all very brilliant at their maxima. The 1006 event reached an intensity about equal to that of the quarter Moon (this at a distance of several thousand light years), and it was almost certainly the brightest celestial sight beheld in historical times after the Sun and the Moon. The 1054 and 1572 supernovae were also long visible in the daytime sky. The remains of the event of 1054 are easily seen today as the Crab Nebula, a glowing cloud of gas located in Taurus and still expanding more than 900 years since light from the original catastrophe first reached the Earth.

No others have been found in the Milky Way since 1604, but in 1987 a supernova explosion went off in the Large Magellanic Cloud, the larger of two of our satellite galaxies in the deep southern sky. At a maximum of the fourth magnitude, it was by far the most recent supernova event that was visible to the naked eye. The Large and Small Magellanic Clouds are two of the nearest separate galaxies, although both are very close to us as galaxies go, and are probably in orbit around our Milky Way. They lie about 160,000 and 180,000 light years from us, respectively; hence we see them as they were about the time that *Homo sapiens* acquired a large brain, succeeding *Homo erectus* by doubling its size around 200,000 B.C., a million years after succeeding *Homo habilis,* a yet smaller-brained creature.

Supernovae are events where the very heavy elements, heavier than iron, are manufactured from lighter elements, and their blowups help to spew the heavier ones back into space, from which future generations of stars still richer in the heavy elements will form. As Carl Sagan has maintained, we are all made of this cosmic stuff composed of most of the elements cooked inside the guts and remnants of supernovae.

Rare though they are, we can see many supernovae every year in the larger telescopes, owing to their brilliance. Most of them come close at their maxima to outshining the whole galaxy in which they reside, and can be spotted across a fair part of the universe. So what if one blew up within, say, 50 light years of the solar system? This will not happen anytime soon because the supernova candidate closest to us, the very bright star Canopus, is about 310 light years away, and it will take perhaps millions of years for Canopus or any other supergiant to approach the solar system. Yet should one occur within 50 light years of the solar system, we are in trouble. In addition to all that visible light, these spectacles also emit loads of lethal energy in the ultraviolet region, causing a deadly influx here quite sufficient to destroy the protective ozone layer in our stratosphere and burn us with radiation, causing widespread destruction and desolation. Those who survived would experience such miseries as melanoma and cataracts in abundance for many years to come.

What causes this to occur? During most of a star's lifetime, the forces inside it are in balance. Gravity seeks to collapse the gaseous nonrigid star while its internal nuclear furnace drives energy outward. A point is reached in the core of a star, where the fusion process occurs, when it is exhausted of its hydrogen, and things become unstable. In stars of moderate size and mass like the Sun, the developing instability causes the core, now mostly helium, to contract while the rest of the star, the envelope, expands, swelling the star into a red giant, after which it may eject mass in the form of successive shells, each appearing as a nova. Eventually the envelope is largely gone and the remains of the core form a white dwarf, a star perhaps only as large as the Earth but with the mass of the Sun. The contents of a white dwarf, or degenerate star, consist of atomic nuclei jammed together, forming material of incredible density; a teaspoon of this matter might weigh many tons. The Sun is likely to end its life in this manner, starting about 5 billion years from now.

Only massive stars, the ones that shine as supergiants, much larger and brighter than normal giants, can become supernovae. They are just too big to collapse fitfully, as will the Sun, and their great mass drives them into a state of gravitational collapse all at once. Actually, a bit of the star may remain in the form of a neutron star, much smaller and denser than even a white dwarf. But the rest will issue forth to enrich the galaxy with the heavy elements that form in trace amounts but in many cases are nonetheless essential to human existence.

16

THE PLEIADES
STAR CLUSTER IS AS CLOSE
AS THE HYADES

Stars are frequently bunched together into clusters; a cluster consists of stars gravitationally bound together, moving with a common parallel motion, and presumed to share a common origin. Clusters come in three different flavors, open clusters, globular clusters, and associations, with associations sometimes grouped together with the more prevalent open clusters into a single class.

About a thousand open clusters are known and recognized as such in our galaxy. By recognized I mean that the cluster contains sufficient stars with a common motion to pass for a cluster and does not consist of just a few stars lying here and there. This is but a small fraction of the total number in the Milky Way because some are not very bright and the interstellar gas and dust in the way does not allow us to detect those that are more than a few thousand light years away. This contrasts with the globular clusters. At latest count we know of 146 of them in our galaxy, with perhaps only another dozen unknown or unrecognized. Globular clusters are so bright and distinctive that they can be identified all over our galaxy, and even in the nearer extragalactic entities, mostly other galaxies.

The best known and most easily recognized open cluster in the heavens is the group known as the Pleiades. To the eye this cluster appears as a tightly knit small group of stars appearing in the autumn sky and covering an angular region a little larger than the full Moon. Known also as the seven sisters, they are recognized in Greek mythology as a group of young ladies standing on the back of Taurus, the bull. The sharp-eyed among us can see more—eight, nine, and even ten separate stars. Binoculars will reveal about twenty more, and a full count of all the member stars would number well into the hundreds.

Older eyes or those affected by cataracts or astigmatism may see fewer than seven stars in the Pleiades, five or six, or with lower resolution, just a faint oval patch in the sky a little larger than the Moon.

Open star clusters come in a wide variety of compactness; it may come as a surprise to some that the nearest cluster to us is the Ursa Major cluster, which includes five of the seven bright stars that form the Big Dipper. Only the two stars at either end of the Dipper are nonmembers, field stars that happen to be in the general line of sight of the cluster. Most stars in open clusters do not cohere for very long, and the clusters are thus not recognized as such for long, even though the member stars themselves may last for a much longer period of time. The member stars are gravitationally linked for a while, but then tidal forces imposed by the Milky Way galaxy and close encounters with occasional nonmember stars and clouds of interstellar gas and dust loosen the group until each star begins to assume its own path through the galaxy. Gradually the stars drift apart; for a while their proper motions are parallel or nearly so and they can be recognized as a group from their common proper motion. But in time even this telltale sign will not hold, and the individual stars will no longer be associated with the cluster in which they were formed.

We speak of the space velocities of stars in two different contexts. True motions in space require knowledge of a star's distance, proper motion, and radial velocity. The distance is found from the parallax, or if that is not measured, we compare the star to one that has a similar spectrum or color and whose distance is well established. The radial velocity is also found from the spectrum by noting the positions of the spectral lines that fingerprint each star. The lines may be displaced toward the red or the blue, toward longer or shorter wavelengths, just due to the radial velocity, the velocity along the line of sight from here to the star. Starlight is shifted in the same manner that sound is shifted. Called the Doppler effect in both cases, it varies the pitch of the sound or, in the case of light, the wavelengths in the spectrum. Whenever a passing car blows its horn, the pitch drops because the sound waves get bunched toward a higher tone when approaching and a lower tone when receding. Similarly, light from an approaching star appears slightly bluer than it would at rest, and redder from a star receding. If one approaches a red traffic signal it will appear green, but only if one travels at a substantial portion of the speed of light, a velocity far beyond any speed limit posted anywhere.

Most nearby stars that fill our skies move about in space at speeds relative to each other of about 10–30 km/sec, as mentioned above. But all of them together are revolving about the massive center of our galaxy at a speed of 220 km/sec. The smaller speeds are just the variations in the motions of the stars, much as each gnat has a slightly different velocity from the average of its swarm. The galactic center is about 30,000 light years from us in the direction of the southerly constellations Scorpius and Sagittarius. Each star or solar system revolves around it with a period of 200 to 250 million years. Thus the Sun with the planets in tow has orbited round the center about 20 times since the system was formed.

Clusters of stars move around the center in lockstep for a while, but then things get loosened up to the point where the stars drift away from each other, spreading the onetime cluster into just another unconnected group of stars. To illustrate this I need only mention that Sirius is a member of the Ursa Major cluster. It has drifted away from the Big Dipper stars to the point that we can recognize its membership only by determining its space motion and finding it to be parallel to the other stars. Sirius appears so bright because it is only 9 light years away, whereas the Dipper stars are about equal to it in intrinsic luminosity, but they appear fainter because they are about 60 light years from us. In another hundred million years or so, none of the stars will be near each other, and the cluster will be recognizable as such no longer.

The Pleiades are an entirely different affair. They are young stars, less than 80 million years old, and they are still tightly bound together, lying at a distance of around 450 light years. Near them in the sky is another cluster called the Hyades. Both clusters form part of the bright constellation Taurus, the bull, but the Hyades are much closer, being less than 150 light years away. The Hyades are seen right around Aldebaran, the bright first-magnitude star of Taurus, but Aldebaran is not a member of the Hyades; it just happens to be located in the same direction at only half their distance, a foreground star.

Even though the Hyades are only a third as far as the Pleiades, they don't shine as brightly. This is because they are much older stars; like the stars of the Big Dipper, they are over half a billion years in age. In Chapter 2, we noticed that the main sequence as it is portrayed in the HR diagram is, among other features, a sequence in longevity. Bright stars near the bright end of the sequence live for only a few million years. Clusters can be ranked in order of age simply by noting the extent of their main sequences; the brighter the top of the main sequence, the younger the cluster. The sequence for the Hyades extends only up to an absolute magnitude of zero, although four red giants also belong to the Hyades. The brightest stars of the much younger Pleiades extend its sequence up to an absolute magnitude of -3 since stars of this brilliance can endure over the shorter lifetime of this cluster. Finally, a still younger cluster in Perseus known as h and x Persei has a main sequence rising into the supergiant region. These are two clusters lying side by side together in Perseus near Cassiopeia and just visible to the naked eye. They are two of the largest and most massive open clusters known, lying between 7,000 and 8,000 light years off. Being less than 10 million years old, stars as bright as absolute magnitude -8 are among their members. Astronomers can age-rank a cluster if color and magnitude information is obtained for at least its brighter members; then one can fit its main sequence to those of well-studied clusters and the nearby field stars and obtain its distance as well.

With this in mind, how would the Pleiades be expected to look in another half billion years? The very luminous stars would have disappeared and the cluster would look no more spectacular than the Hyades or the Ursa Major cluster, both of which are about half a billion years old. In fact the Pleiades would appear much fainter and less spectacular since they are much farther away. What happens, then, if the distances of the Hyades and Pleiades in their present states are reversed?

The brightest stars in both clusters appear at about the third magnitude but if switched, the Hyades would be visible only in binoculars while the Pleiades would dominate Taurus and the entire autumn sky. That one small group would contain four stars of the first magnitude and one of zero magnitude. Bring them yet closer, to the distance of the Big Dipper stars, and they would sport no less than five stars brighter than any other but Sirius, all in an area smaller than the bowl of the Dipper. They would be worshipped as gods on a celestial Mount Olympus or Valhalla of their own.

Associations are considered special cases among open clusters. They are young groups composed of young stars. This means that many of their stars are very luminous and very hot, making them blue stars. If we look at Orion, we find that six of his brightest seven stars that form his shoulders, knees, and belt are all blue supergiants moving in a parallel way that reveals a common origin. Only red Betelgeuse is of a different age and origin. Many of the member stars, in particular the fainter ones making up his sword, are extremely young as stars go; in fact, a glance at Orion shows that one of the three brightest of the stars that make up his sword appears fuzzy, even to the naked eye. Stars embedded within this great assemblage known as the Orion Nebula are the youngest of all—it is their excess ultraviolet radiation that excites the surrounding luminosity into a blob of mostly hydrogen gas that glows in the reddish region of the spectrum. The Orion Association comprises all of the nebula as well as the other bright stars in that brightest of star figures.

Associations appear as collections of bright blue stars spread out over a considerable expanse of the sky. One other association dominates its celestial region in addition to the Orion Association. Opposite in the heavens to Orion is Scorpius, as mentioned earlier. Most of the stars that make up this creature are also bright, blue, young supergiants. Again like Orion, one of the brightest stars is the red Antares. This group covers a larger area to the south and east of Scorpius, including the small constellation of Lupus, the wolf, as well as Centaurus, one of the largest and brightest constellations in the whole sky. Centaurus we know from its brightest star the nearby Alpha Centauri. But Beta Centauri just 4 degrees to the right, as we see it, is also bright. Not far away lies Crux, the famous Southern Cross, with two more first-magnitude stars and a continuation of this giant association, the Scorpius-Centaurus Association, sometimes abbreviated to Sco-Cen after the two largest constellations covered by its stars.

Orion and Sco-Cen are our two brightest and nearest star factories, places where stars are being born even today out of the surrounding interstellar gas and dust. If either of these sprawling groups happened to be not a thousand but only one or two hundred light years away, we would all be inside the volume that they fill, just as we are within the extent covered by the Ursa Major cluster, as shown by the locations of Sirius and the Big Dipper stars.

The proximity of a cluster or association of stars to us, suggested in this chapter, might not sway our celestial heritage from what it was into something else as much as the rearrangements of the Sun, Moon, and planets would surely have done, unless by happening to lie very close to us, a group's brightest members became visible in the daytime in full sunlight. Then the effect would have influenced astronomical history, not so much in classical times, when observers knew the stars didn't "go" anywhere but remained in the sky, though invisible in the daytime. But in earlier societies this was not widely known—stars were thought to be eaten by dragons or befall some other diurnal malady at dawn, only to be magically resurrected at dusk. If some of them were with our ancestors and a part of their world throughout the circadian pattern, day as well as night, I suspect that some of those early cosmologies would be very different.

17

THE GREAT
POPCORN BALLS

The southern end of the city and island of Miami Beach, known as South Beach, occupies the southern tip of that municipality and island. South Beach has graduated from a region of modest housing for older middle class people to an upscale neighborhood for swinging singles and others. After sunset, with palm trees swaying and clicking in the breeze in the balmy twilight, the city becomes awash in light, but here in South Beach it stays dark, devoid of outdoor lights, forbidden by municipal ordinance for a reason that becomes obvious as the dusky sky comes alive and the first stars emerge from their daytime invisibility. Low in the southern twilight sky shines a nearly circular swarm of stars bulging very slightly out from a circle with five nearly symmetric lobes, hence its name of Cinquefoil. No larger in angular size than a fist held at arm's length, this gaggle of first-magnitude stars appears to grow in the darkening sky as fainter stars come into view.

Later as the sky darkens, one can no longer cover the group with a fist, nor even with an extended hand. The circle of stars widens and deepens until in the full darkness it grows into a beehive of hundreds of stars, the brighter ones as red as Mars, glittering and shining in the black sky. And the twinkling! Never has stellar scintillation been more vivid and pronounced than it is on these winter evenings when this cluster is at its highest, due south in the heavens. Ever mindful of the swarms of snowbirds who migrate here from the north in the wintertime and keep Florida's economy in a bullish state, the city had doused its light pollution at this southern tip so that they can come and watch Cinquefoil unfold in its full glory, itself a major tourist draw, with a prominent likeness appearing on the state flag.

Cinquefoil is a globular cluster, a spherical pack of hundreds of thousands of stars all gravitationally bound together, with the brightest stars bunched toward its center and the fainter ones filling in around the edges. Unlike a hundred others in our Milky Way galaxy, this globular cluster is by chance extraordinarily close to us. All but invisible in the northerly latitudes of New York and Chicago and almost all of Europe, it is here a spec-

tacle, a symbol for the entire state of Florida, gleaming all winter long in the southern sky.

Most telescopes belonging to amateur astronomers have apertures of 6 inches or less and are portable in the sense that they can be carried out to a clear spot and stowed when not in use. A few are larger, up to 10 inches, and in rare cases, as much as 20 inches in aperture. Universities and planetaria are more likely to own a telescope of aperture 12 to 24 inches, for which some kind of dome or sliding roof is customary. Astronomers almost always refer to a telescope by its aperture, defined by the diameter of the largest optical component in its optical system, be it a lens in the case of a refracting telescope, or a mirror in a reflector.

As one might expect, the cost of telescopes increases greatly with an increase in size, by which we mean aperture. A century ago, most observatories owned refractors, but now reflectors are made of comparable optical quality and are much cheaper to buy and mount, and with their much shorter tubes, the diameter of the dome enclosing it can be smaller and less costly.

It may be surprising to learn that for the casual or first-time observer, the 6-inch telescope shows most celestial objects just about as well as does a larger instrument. Sky brightness due to light pollution is the primary limiting factor for most faint nebulae and other sights, hence an object like the Orion Nebula or the galaxy in Andromeda reveal little more when glimpsed through the larger aperture. One noteworthy exception is the globular cluster. The reason for this is that the 6-inch telescope can rarely resolve the individual member stars in even the closest globular clusters, whereas a 24-inch instrument can and does so regularly. What appears as a fuzzy star in a 6-inch telescope is seen in one of 24 inches or more to resemble nothing so much as a great popcorn ball (see Fig. 17.1).

This is a worthy sight on a good dark night. The stars in the outer regions of the cluster appear in the thousands; in fact, careful counts have been made of cluster stars that number as many as 30,000 in a single cluster. In actuality, the central and faint stars cannot be resolved, and by mass estimates we calculate that several hundred thousand stars should be a more realistic estimate—this in a volume of space perhaps 50 to 100 light years in diameter. We first consider the night sky view from an Earthlike planet near the core of the cluster where the star density is greatest. What would we see?

Fig. 17.1 View of a very nearly globular cluster as seen from the Earth. Painting by Chesley Bonestell. Used with permission of Bonestell Space Art.

Here we pause to introduce the one remaining distance unit astronomers use. For distances within a solar system, the 93-million-mile (150-million-kilometer) astronomical unit is most convenient in order to spare us the overuse of millions and even billions of miles or kilometers. For interstellar use, the light year is better; one light year equals about 63,240 AU or nearly 6 trillion miles or 10 trillion kilometers. But because stellar distances were first determined from the sizes of their measured parallaxes, always in units of arc seconds, the parsec (shortened from the words parallax and second) is preferred, where the distance of a star equals the reciprocal of its parallax and thus also, as it happens, in terms of 206,265 AU. Anything seen at 206,265 times its own diameter will appear as one arc second in diameter in angular measure because that is the ratio between a second of arc and a radian (or radius of the sphere with the star at its edge). Since the astronomical unit is now known to be 149,597,870 kilometers or 92,955,807 miles with an uncertainty of only one mile, the parsec is as tightly defined in miles or kilometers as is the light year. One parsec turns out to be 3.26 light years or near 30 trillion kilometers or around 20 trillion miles. For yet larger distances we just add the metric prefixes; thus, kiloparsec and megaparsec are always defined as 1,000 and 1,000,000 parsecs, respectively, and when one speaks of a cubic parsec as a unit of volume in space, you know this is the same volume as occurs within a cube 3.26 light years on a side.

The stars in a globular cluster, several hundred thousand of them, must fit into a volume only maybe 50 parsecs in diameter. This is nearby space to a stellar astronomer. Naturally, any assemblage of objects held together only by their mutual gravitation fits a density law that is highest at the core and lowest at or near the edges, however defined. The numbers rise from about 0.4 stars per cubic parsec at the edge to between 100 and 1,000 at the center or core. Compare this to the star density in the solar neighborhood, defined here as the space within 5 parsecs of the solar system. This volume amounts to about 524 cubic parsecs, and we know of 67 individual stars in 51 double and multiple star systems; that is, a star system like Alpha Centauri counts as three stars in a single system. The stellar density near the Sun is hardly more than 0.1 stars per cubic parsec; therefore the stellar density in the cluster ranges from about 4 times ours at the edge to 1 to 10,000 times at the core. Despite such high stellar densities, the chance of a collision between one star and another even at the core is very rare; one indication of this is the fact that if a straight line were passed through the cluster at its core, the chance of hitting even one star is less than 1 in 100 billion!

To be able to imagine what a night sky would look like from a planet revolving around a sun in the core of a globular cluster, we must convert these densities into the numbers of stars visible and brighter than any given magnitude. This is not easy to do because the frequency of stars of differing luminosities is not the same as it is in the solar neighborhood (for example, about a third of our brightest stars are supergiants, whereas few if any supergiants, and hardly any blue ones, are found in globular clusters). Nonetheless if we assume a number closer to the low end of the range, we may not be far off in what the night sky would look like from a planet at the core. Here at our present location we see two stars at magnitude −1, eight at 0, and twelve at +1, all magnitudes including the interval between 0.5 magnitude brighter and fainter than the whole magnitude given. Thereafter as we proceed to fainter magnitude intervals, we find about 60 stars of the second magnitude and 200 of the third and for each successive whole magnitude the numbers triple. For at least a general idea we can simply multiply the numbers by 2,000; thus one should see in the entire celestial sphere 30,000 stars at +1 magnitude, 16,000 at 0, and 4,000 at −1. From this we might sensibly conclude that we would also have at least a few hundred at −2 and perhaps a hundred at −3 and a handful even brighter. Imagine a sky seen from a planet's surface (half of the entire sky) with more than 10,000 stars of the first magnitude or brighter, with several times that number as bright as Polaris and the stars of the Big Dipper, all visible from one place on our globe. The whole sky would shine every night with the brightness of our sky when the full Moon is a part of it.

All globular clusters are old; they were formed early on in our galactic history, more than 10 billion years ago, up to about 13 billion years ago—since that is only a little younger that the presumed age of 13.7 billion years for the universe, they cannot be older than that, no matter what cosmology is assumed. The Milky Way, like all galaxies and solar systems, started out as a nebula roughly spherical in shape. Then as it collapsed due to the mutual attraction of its contents, it began to rotate around an axis. But in the plane of its equator the centrifugal force balanced the collapse and prevented further contraction. In other directions the collapse continued until a thin disk was formed. Just as most of the planets lie in about the same plane, so do most of the stars lie in a thin disk. But somewhere before that collapse, many of the globular clusters came into being. Unlike stars and open clusters, these old clusters retain, collectively, a spherical distribution, a kind of round halo around the rest of the Milky Way. With so many stars in a small volume in the globular cluster, they retain their members through mutual gravitation, in the cluster form, throughout their life spans. They are thought to be older than the other stuff of which the disk is formed, and at the time of their formation we know that there were very few elements heavier than helium around. After all, these heavier elements formed in the interiors of stars, particularly the very bright, massive stars whose supernova explosions enriched the interstellar medium with the heavy elements they had produced. Then, as later stars were formed, they incorporated the heavy substances into themselves and their planets.

Before the First World War it was assumed that our Milky Way was a flattened group of stars centered near the Sun and falling off in density slowly with distance in its plane and much more rapidly in directions away from this disk. Sir William Herschel was a pioneer in counting stars at each magnitude and in every direction, and he came up with this simple model. What Herschel could not know is that interstellar gas and dust abounds in space, particularly along this galactic plane, the plane of the disk. Then during the war Harlow Shapley, one of the great astronomers of the twentieth century, derived the distances to the globular clusters and deduced that they followed a density distribution roughly spherical in shape and centered not here near the Sun, but some 30,000 light years off in the direction of Sagittarius, deep in the southern summer sky. Shapley had gone Copernicus one better. Copernicus and his followers proposed and later confirmed the heliocentric model for the solar system, causing endless disturbances and challenges to the theology of the day. Shapley once again moved the center of things away, this time from one AU to over a billion. Perhaps the religions and philosophers had learned a thing or two in the past three centuries, for although this made the front page of the *New York Times* more than once, it repeated the earlier debacle not at all.

It is then very unlikely that a planet like the Earth and its life forms would have formed as part of a globular cluster, owing to their much greater age and lower metal content. If we want to imagine ourselves at the center of such a rich old accumulation, it would be much more likely that we would do so as a much younger solar system that happens to orbit through a globular cluster as an interloper and not as a true member of the group.

We must take note of one other probability. As the solar system passes through the core of the cluster, it is likely that we would pass very close to one or more of its stars, quite possibly close enough for their gravitation to perturb one or more of our outer planets into orbits no longer bound to the Sun. Or worse, an outer planet might be propelled into an orbit bringing it inward near the Earth with all the dangers that implies.

How do we know that all globular clusters are very old, among the oldest objects in the Milky Way? From spectroscopic and other studies we can assign an index of the ratio of the abundances of metals such as iron in the cluster members to that in the Sun. In the Sun the abundance of Z, the percentage of all elements other than hydrogen and helium, accounts for only around 2 percent of its total mass. As we have seen, these heavier elements came largely from the remains of the supergiants, which produce them in their interiors. In the supernovae these heavier elements get spewed out into the galactic medium, enriching it in the process. Thus, the amount of Z, including iron and the metals, has increased with time from near zero to a couple of percent—perhaps not at a constant rate but increasing all the while. The globular clusters, being old, were formed when few metals were made and so are limited mostly to hydrogen and helium. Astronomers express the metal content of a star by $[M/H]$, which is defined as the logarithm of the ratio (M/H) for the star divided by (M/H) for the Sun; hence, a value of zero means that the solar abundance applies, whereas -1 means that the star has one-tenth the metal abundance, and -2 means that it contains only one-hundredth the abundance for the Sun. The range in $[M/H]$ for nearby stars in our neighborhood varies from zero to about -0.7, or from the solar abundance to one-fifth that amount, with a few exceptions. Globular cluster stars, however, range from about -0.5 to -2 or less; they are metal-poor, and from this and other data we know them to be among the galaxy's senior citizens. As such they have no bright blue stars; all the brightest would be red, the color of Betelgeuse, Antares, and Mars.

The spectacle of a nearby globular cluster would be a handsome sight. We would see this old-timer with its bright red stars filling much of an entire hemisphere of the sky. From within we would be less aware of its overall shape and appearance, but the sky would be starry to a level we can only imagine. We have discovered and catalogued only 146 of these round stellar agglomerations in our galaxy, and there cannot be even a quarter of that number yet waiting to be found. They are bright and they can be seen throughout the Milky Way unless they happen to lie behind the giant galactic core or a dusty spiral arm; many are visible in other galaxies millions of light years away. They all possess a number of common characteristics and are reasonably of a size—variations exist, of course, and a researcher would rarely confuse one with another, just as a stable manager would not confuse his horses. At a glance, however, the casual observer can tend to confuse them and register the very pejorative rejoinder that if you have seen one, you have seen them all.

18

THE MILKY WAY LIES
ALONG OUR EQUATOR

Coordinate systems are devised to identify uniquely each and every point on a surface. They typically fix points on a plane, a left-right measure commonly labeled the x-coordinate and an up-down scale known as the y-coordinate. These kinds of systems are called rectangular or, more commonly, Cartesian coordinate systems, named after René Descartes (1596–1650), the eminent French philosopher and mathematician. I suppose the best-known Cartesian system of coordinates is the following. The levels of the y measure might be labeled 4, 14, 23, 34, 42, 50, 57, etc., and the x lines might be 1, 2, 3, etc., or, more specifically, 1, 2, 3, Lexington, Park, Madison, 5, 6, etc. Yes, this is the coordinate system designating the streets and avenues of New York City or, more specifically, Manhattan Island north of the Village. Any New Yorker recognizes the y-divisions listed above as the major stops along the IRT and BMT subway lines running in the north-south directions. Each point in the scheme is designated by coordinates (x,y) so that (5,42) uniquely describes Fifth Avenue and Forty-second Street. We ignore Broadway for the nonce, slicing as it does through the system at an oblique angle like a rampant bolt of lightning.

Most systems have an origin described by (0,0) but in New York's case, one would have to move a block to the east of First Avenue and four blocks south of Fourth Street to find it. Such a point exists, although it has no significance to most New Yorkers since it lies somewhere between Houston and Delancey Streets a little east of the Bowery! Admittedly, the Manhattan coordinate system, like those of most cities of Cartesian design, doesn't always maintain equal spacing between one block and another. In general, however, the pattern is spaced 6 blocks to the mile between the numbered avenues and 20 blocks to the mile in the north-south layout of the streets. Thus the Empire State Building at (5,34) and the Chrysler Building at (3½,42 — Lexington Avenue being midway between Third and Fourth [Park]

Avenues) are separated by the hypotenuse of the triangle with legs of $1\frac{1}{2}x$ (1/6) + $8x$ (1/20) miles; this amounts to sqrt ($0.25^2 + 0.4^2$) = 0.47 miles; hence, the separation between these two tallest towers is just under one-half mile.

The other common type is the spherical coordinate system, a pattern used to designate points on a grid wrapped around a globe. To be sure, New York is part of the surface of a sphere, but it is a sufficiently small part of it that a plane representing the city is adequate. In the case of a ball, a fundamental great circle exists (a great circle is defined as a circle on a sphere whose center is coincidental to that of the sphere—it is also the largest possible circle that can be fit onto the sphere). The Earth's equator is such a great circle, the most fundamental of all. Any great circle with an axis perpendicular to it passes through the north pole and the south pole. The axis connecting the poles is the axis of rotation of the Earth, and is the obvious choice for the fundamental reference frame on its surface. For centuries navigators have defined latitude as the angular north-south measure from the equator defined as the circle of latitude 0 and the poles as 90N and 90S, or +90 and −90, for the north and south poles, respectively. A point midway between the north pole and the equator lies in latitude 45N, and so forth.

Now it is clear that we need a reference point for east-west measure, known as longitude, but no clear unequivocal point exists in nature, and we are required to define one arbitrarily. Actually, the requisite reference is not a point but a half great circle extending from one pole to the other, called a meridian. Any meridian will do, and seafaring nations established their own meridians, ones that usually passed through their capital cities. Thus France defined one that passed through the Observatoire de Paris and the Jardin du Luxembourg; Great Britain's went through the Royal Observatory in Greenwich, an eastern suburb of London; and that of the United States passed along Sixteenth Street in Washington, D.C., and through a park appropriately named Meridian Park. Since observatories had the best instruments and observations for precise longitude measurement, they were the best places for a zero point in longitude. As more countries made up more meridians, confusion inevitably mounted, and furthermore time became defined in a way that a specific amount of time became equivalent to a slice of the Earth that rotates into itself in just that amount of time. When travel was slow, whether by oxcart, stagecoach, or clipper ship, the different times in each city conformed to the angle between its meridian and whichever one served the nation involved. Thus New

York's local time was 12 minutes ahead of Washington's, and Chicago's was 40 minutes behind. With railroads, land was covered fast enough to make an absolute mess of railway timetables, confusing enough to read without the almost infinite number of time systems that could be defined. A train proceeding from Boston to Washington moved swiftly enough to require a traveler to reset his timepiece 12 minutes behind as he approached New York, another 5 minutes slower for Philadelphia, and a further 7 minutes to agree with Washington's local time when he arrived there. Railroad schedules were difficult enough to deduce without this complication added on.

The American railroads themselves came up with a solution to this untimely disorder in 1880, and Congress had the sense to approve it. This was the creation of time zones; each zone included an hour's worth of east-west measure, known as longitude, with deviations allowed in order to conform to state lines and national boundaries. Thus nearly the entire east coast had the same time throughout the states that comprised it. In 1884 an international convention was held in Washington in order to convince all nations to settle on one and the same prime meridian for the whole world. After the consideration of many points, including Rome, Jerusalem, and the Great Pyramid, they chose the British meridian of Greenwich for the world. It was superbly defined by the Royal Observatory there, and besides, Britain had the most powerful navy.

Since the time of the classical civilizations we have known that the celestial sphere, as its name implies, can be portrayed as a sphere on which all the stars are pasted at a distance of infinity for most purposes. The Copernican system destroyed the reality of the concept, but it is still useful for navigation and the definition of a coordinate system for the sky. The simplest arrangement for positions in the sky arises from the extension of the Earth grid onto the sky. We extend the geographical axis onto the sky; the poles intersect the celestial sphere at the points labeled the north and south celestial poles, and the celestial equator is formed by the projection of our equator onto the heavens.

However, it is worth noting that two other celestial systems are apparent; one arises from the ecliptic, itself a great circle, and its two poles, the north and south poles of the ecliptic. These are by definition $23\frac{1}{2}°$ from the celestial poles, just as the ecliptic and the equator are inclined also by a $23\frac{1}{2}°$ angle. Finally, the thin galactic disk, which we see as a faint glowing band girdling the heavens, can be defined with an equator through its middle and two poles at the usual right angles from it. This is the coordinate system that defines the orientation and the rotation of the galaxy, and it is tilted a whopping 62.6° to the celestial equator. Astronomers have examined the possibility of an alignment between the plane of the galaxy and the planes of the double stars within it and found none. The plane of our solar system and the double and multiple star systems appear randomly distributed. The high tilt of the galactic plane to that of our system shows no disposition for alignment between the two planes. The Milky Way courses up north through Cassiopeia, a northerly constellation, then descends dramatically through the summer sky past Scorpius to the sprawling constellation of Centaurus and its compact neighbor, Crux, the brilliant Southern Cross in the deep southern sky, invisible north of southern Florida.

On a dark, clear moonless night in midwinter and again in midsummer, the visible stars show a preference for the thin band of the Milky Way. A few very bright stars seem to vitiate this tendency, no star more than Arcturus in the spring sky, and the stars forming the nucleus of the Big Dipper cluster to its north. Remove these and the spring sky is all but devoid of bright stars. Not until the embodiment of the summer heavens, the "summer triangle" marked by Vega, Deneb, and Altair, rise in the east do the attendant second- and third-magnitude stars again fill the corners of our sky.

Similarly, the stars of autumn, filled with their representations of ancient legends, are altogether a muted lot, blandly forming star patterns of note but subdued in luminance. Try, for example, to view the sky at a time when the singular Pleiades transit the upper meridian and are situated due south as close to the zenith as they can get. This occurs in midevening in the late fall or with twilight and approaching darkness in midwinter. Look for them high in the southern sky on a dark clear night, of course, and look to the east and to the west. To the west of this unmistakable cluster, where the autumn stars are fleeing to the western horizon, the sky is bleak. But east of these seven sisters a plethora of luminaries has arisen. Orion and his two dogs, Canis Major and Canis Minor, the former with Sirius, the lucida of the entire sky, gleaming and twinkling its steely blue color, and the latter with white Procyon, dominate over lesser but nevertheless bright stars forming the twins, Castor and Pollux, and to their north yellow Capella. A mix of happenstance and the dominant galactic disk illuminate these frigid night skies, equaled only by the deep southern star groups of Centaurus, Crux, and the sprawling Argo Navis, one of Ptolemy's constellations, so vast that in modern times it is divided into four parts of the ship that carried Jason and his Argonauts in search of the Golden Fleece. Now the sails, the keel, the stern, and the compass each delineate their own smaller but brilliant domains, with Canopus blazing over the whole.

What would we see if the Milky Way were coincident with the equator and not skewed at a jaunty 62-degree angle from it? From our mid-latitudes, a band of hazy light would be omnipresent, rising directly out of the east and the west and merging at a high point in the south while the northern sky appears undistinguished. From northern Europe or Alaska, the band of light would be mixed with the haze in the lower sky from light pollution and natural causes.

We can amplify the discord by supposing ourselves to be well outside and to the north of the Milky Way, which would then lie well to our south. From our position outside the conglomerate we could view the vast, gorgeous, spiral nature of the whole thing, the spiral arms that are now hidden by way of our being embedded within and between them. The northern sky appears black and undeniably empty—empty of planets for they lie along the ecliptic, and mostly devoid of stars as well. We know from the motions of our fellow stars in nearby space that very few if any stars are moving fast enough to escape from the galactic mix due to its massive gravitational potential binding them to the system. No, the northerly heavens are totally dark unless an auroral display or the telltale presence of urban glow from an excess of streetlights intrudes against the blackness.

To the south we see the framework of the Milky Way, its arms outstretched across half the nocturnal sky. Its center or core forms the most brilliant thing up there after the Moon, casting shadows throughout the night, although being to our south it could not be seen in its full splendor much farther north than the equator. The core or nucleus appears reddish in sharp contrast to the spiral arms dominated by the blue stars, brightest of all, aligned along the arms and accompanied by knotty blobs of nebulosity. An occasional fuzzy globular cluster appears here and there, but most clusters of this kind would crowd up to the center, and the rest are scattered in all directions. The mighty synergy of this rotating wheel would so dominate the celestial realm that it would certainly be worshipped as a god among gods. Like the lights of a city viewed from a nearby prospect overlooking it, the supergiants would gleam and sparkle like distant streetlights; most appear blue, with a red one now and then to break the uniformity.

We assume that we are to the north (above) the center of the galaxy by some 50,000 light years, a distance about equal to its radius. From our actual perspective in the plane we see one spiral arm in front of another and all corrupted by interstellar gas and dust blocking the view we would otherwise have. But from above we see the whole picture, spanning some 90 degrees across from one edge to the other. The center casts shadows; it is as bright as Venus, and the spiral features are not much fainter. Since we know few if any stars escape its gravitation, we have no stars in our northern sky. We might just see a fuzzy round patch or two that prove to be globular clusters, but no other object is visible to the eye save our own planetary system.

19

WE ARE ALONE II

Ours is not the only galaxy; there are many others, billions altogether, scattered throughout the visible universe. Three of the nearest among them have been mentioned above; two of these are the Large and Small Magellanic Clouds, the two small satellite galaxies to our own Milky Way, about 180,000 light years from us. The third is the great spiral galaxy in Andromeda known also as Messier 31, or M31, a near twin to our own Milky Way in size and majesty. Lying almost 2 million light years off, it is the farthest object visible to the naked eye.

Charles Messier was a French astronomer and comet hunter of the late eighteenth century, who became annoyed at the permanent fuzzy objects that were scattered across the sky and could occasionally be mistaken for comets. He took the time to list the most conspicuous of these fuzzy objects; his list extended to 110 objects of many different kinds but all sharing one property, that of revealing a fuzzy image when spotted through a telescope. Some turned out to be star clusters, some were galaxies, and some were nebulae, irregular blotches of interstellar gas shining by several different processes. The ones with highly descriptive names, such as the Ring, Dumbbell, Crab, and Horsehead Nebulae, are almost all on Messier's list.

One of the common kinds of nebulae is colored with a deep crimson red hue and another shines with a vivid blue tint. Those of the first kind are called emission nebulae, whose red color results from stimulation by nearby hot, young blue stars emitting powerful ultraviolet radiation, sometimes embedded within the nebular material. These emissions stimulate the surrounding gas, mostly hydrogen, to glow with a red hue characteristic of hydrogen gas.

The blue nebulae are reflection nebulae and shine by simple reflected light commonly from the same types of hot blue stars. The great Orion Nebula, the apparently brightest of all gaseous matter and visible to the naked eye, as well as the diffuse matter surrounding the young Pleiades star cluster, are typical of emission (red) and reflection (blue) nebulae, respectively. The round, greenish planetary nebulae are so named because they are round and resemble the two greenish-tinted outer planets Uranus and Neptune. These nebulae are caused by a nova-like explosion of a star central to the nebula. Most of the rest, the so-called spiral nebulae, as they used to be known, are galaxies. The Milky Way and its twin galaxy in Andromeda are both known to be spiral galaxies since they exhibit spiral arms in profusion.

One other galaxy completes the list of island universes visible to the naked eye; this is Messier 33, a spiral galaxy in the small constellation of Triangulum, not far from Andromeda but somewhat smaller and fainter and almost as distant as that larger galaxy. It is visible to the unaided eye but barely; it takes above-average visual acuity and near perfect sky conditions to glimpse it. M31 is the farthest object visible to the naked eye, lying over 2 million light years off, with M33 just a little less distant. These two galaxies, with ours and the Magellanic Clouds, are the five largest members of what has become known as the "local group" of galaxies, a collection of about twenty-two of them that remain gravitationally bound to each other. The Milky Way and M31 are the two big guys in the group, but they are by no means the largest known anywhere. An indication of this is given by the number of globular clusters belonging to a galaxy; ours has between 150 and 200, whereas M87, a great giant elliptical galaxy in the Virgo cluster of galaxies, has more than 4,000 and maybe as many as 10,000 globular star clusters!

The smallest galaxies in our group warrant the name but not by much; some of them are no more than glorified star clusters that are identifiable partly on the basis of their location well away from any other local group member. The galaxies in the local group are distinctive in that they do not appear to share the uniform expansion of the universe first noted by Edwin Hubble, the motion that causes more distant galaxies to recede from us at speeds in proportion to their distances.

Suppose our solar system was positioned not as it is, in the plane of the Milky Way two-thirds of the distance from the center to the edge of the disk, nor about the same distance above or below it, but instead halfway between it and M31, about a million light years from either of them. They are near twins in size and mass, thus both appear as fuzzy patches about a degree in length and of the third magnitude; the two appear the brightest objects in our night sky beyond the solar system.

No star other than the Sun can be seen without a large telescope, and the sky would appear a uniform black. Only when telescopes of modern size became available could we resolve their spiral arms into stars, mostly young, blue supergiants. Being young, these youthful stars have not had the time to drift far from the arms themselves, made of interstellar gas and dust, from which stars are born, and therefore the blue stars align along the arms themselves and give them a soft, tenuous, bluish luminescence they would not otherwise have.

What would we make of such a discovery? Two faint fuzzy objects don't form much of a coordinate system against which we can measure the motions of the planets, just as now we can't perceive the Sun's motion against the invisible stars surrounding it in the daytime. We would by now, with our level of technology, establish a reliable and accurate coordinate system for the heavens, but only in the last few decades or so. Neither Plato nor Aristotle nor any of their followers through even Newton and Halley could detect more than the crudest of planetary motions, and Herschel would have hit upon Uranus only by chance, as he did, but then he would likely fail to note it as a planet with no stars for comparison. Bradley would have no background against which to measure the aberration of starlight, and Bessel would not know the parallaxes or proper motions of stars he had no hope of seeing. The distance scale of the universe would be very poorly known even today—from Pluto to even the closest galaxy is just too great a distance ratio for astronomers to bridge with any or all of their methods of detection.

We miss our stellar system just as we miss our solar system as imagined in Chapter 4. When the Moon and the bright planets are not above the horizon at night we see blackness in every direction. No Big Dipper, Orion, Pleiades, or Cassiopeia breaks the expanse of black. Since the moving objects (Sun, Moon, and planets) move with respect to nothing, their special significance may be greatly reduced in importance. Where would astrology be with no signs or meaningful houses to populate the framework of the horoscope? Birth signs have no meaning and no simulacrum of relevance to our world. Precession may still have been realized by Newton, but it could not be measured until many years later when the mechanics of the solar system were worked out. The astrophysics of the Sun will not be understood with certainty with no other stellar spectra or masses known. The largest telescopes might be much smaller than they are with little to observe unless and until faint objects like Pluto and distant galaxies were found, and it might not have occurred to anyone to carry out the search for them.

Classical mythology, from Homer to Virgil, would have taken on a different cast, and likely a poorer one. With no Perseus, Andromeda, or Pegasus in the sky, might these myths be far more obscure or have disappeared altogether? The same might also hold for the stories of the Chinese, Norse, and Native American cultures, many of which were also tied indissolubly to the heavens.

If both the premises of Chapter 4 and this chapter were combined, we would have only the Sun and the Earth, together making up the entire universe. Nothing else, not even the Moon. By now we would have a good idea of the size and distance and mass of the Sun (as well as the Earth), but what would or could we make of the limitless void beyond? The Sun provides us with the heat and energy we need to thrive, but what a dull night sky we would have—it would be dark throughout the night and the year. Who would care whether the night sky was cloudy or clear? Nothing would be visible in it; all nights would be dark, and on any one of them it might not be possible to tell the difference. No Moon—only our own bonfires and later electric lights would provide the means to see after sunset.

In one way the two-body universe would be much simpler. Aside from the solar gravitational influence on the Earth's equatorial bulge, the motions would be far easier to calculate. The two-body problem in celestial mechanics is a fairly standard one, and although it lacks a rigidly precise solution, the orbit of the Earth would be quite exactly replicated. The bulge effect would yield a modicum of precession but not nearly so much without the participation of the Moon. How could we produce an inertial nonrotating reference frame as we do now from distant stars, more distant galaxies, and the very precise orbits of asteroids? We could not—not with any certainty that the system, such as it is, does not rotate at least a little bit. We would have determined the lengths of the year and the day long ago, the one from repeated arrivals of the Sun at either solstice, when its shadows reach maxima or minima, and the other from the Foucault pendulum and the sidereal and solar days from the apparent motion of the Sun. The mass of the Sun would be much more difficult to ascertain.

Relativity and quantum physics might have been slower coming or stagnated earlier than they did because none of the tests that Einstein and others contrived to affirm or deny their reality would have been possible. He and Hubble might never have achieved the worldwide prominence that came to them, the stellar role in such processes as nucleosynthesis would be missing, and science—big science—might not have been launched as soon or as successfully as in fact it was. This is but one of the alternative histories that could have made the Manhattan Project all but impossible, if not inconceivable, delaying the development of nuclear weapons and allowing war to be framed in less frightening terms for the remainder of the twentieth century. Perhaps World War III between the United States and the Soviet Union might have been fought as a hot war, not a cold one.

IV
HOMEMADE
SKIES

20

RING OF RUBBISH

Artificial satellites have been with us since October 1957, when the Soviet Union launched the first *Sputnik,* almost half a century ago now. Since that time satellites have become more numerous and sophisticated, but in one respect they have not changed. This is their visibility, their luminance from reflected sunlight as they spin about the world. Space is still mostly the province of governments and nonprofit institutions, although the corporate world has shared payloads on any number of them.

Soon this will change. The expense of a launch into near-Earth orbit and the maintenance of a modest payload have dropped by orders of magnitude and will continue to do so—even now the first tourists have gone aboard a space capsule into a near-Earth orbit at $20 million a pop. What happens when the corporations have total access to nearby space; what vanities will they pursue that have not yet been seen in the night sky over the planet? Surely one of the very first great changes will center on advertising. The technology is there now to inflate an enormous balloon of mylar and bobble it into near-Earth orbit only a few hundred miles overhead. Satellites when in the sunshine are bright; in fact, there may be no other place in which day and night light conditions are so closely juxtaposed. They typically shine as luminous tiny moons (which in fact they are), sometimes with the sheen and luster of Venus, the brightest pointlike object in the heavens and the planet with the highest albedo.

A globe appearing larger and much brighter than the full Moon in the sky with a corporate logo or slogan blazing down on us for hours after sunset and again before sunrise is a real probability and in fact has almost happened more than once. To gain an impression of these skies of the future, one has only to look at any strip along a highway. Junk lighting pervades everything in sight except for an occasional dark corner or crevice, to the point where the circadian rhythms of most creatures living nearby are grossly affected. Trees can retain their leaves in this eternal daylight to the limit of survivability. Birds circle around and around illuminated buildings, bridges, and signs to the point of death from collision or exhaustion. Residents nearby suffer through an interrupted melatonin secretion, leading to an increased incidence of cancer and possibly other maladies. If we do that to our planet, we are very probably going to do it to the sky.

On the basis of recent medical research, the International Dark-Sky Association issued a statement in November 2002, parts of which are quoted here:

> Recent research has made it clear that we must take into account that electric lighting affects more than vision. There are numerous photobiological aspects to light, with both positive and adverse impacts on humans and on the ecosystem. These effects of light have grown with the increased use of electric light and they are critically important to our quality of life.
>
> The issues are complex. Research on the photobiological aspects of lighting is now quite active, with many players. It is very important that we all be aware of these issues. Quality and comprehensive research is much needed (just as it is in vision) to gain understanding. There is much we have to learn. Several things are now certain however. As humans, we need both bright days and dark nights: "Healthy lighting, healthy darkness." More light is not necessarily better. Light should be carefully used with thought given to intensity, timing, duration, and color.

Even now space junk orbits the Earth like a ring of filth with thousands of pieces of things identified and tracked by NASA and similar organizations; the only saving grace is the invisibility of most of it to the naked eye. When the advertising balloons flourish forth they could each outshine the full Moon, and near daylight conditions would be perpetuated into the night. What happens to these abominations? Micrometeorites, small grains of rock as fine as sand, pervade the interplanetary realm. Those specks, which can be seen only as shooting stars on a warm summer evening as they streak through our upper atmosphere and burn up in it, are densely distributed in nearby space to the level that they are collectively visible as the zodiacal light, the so-called false dawn that rises before the first hint of the real dawn. These meteors will in time tear the advertising slogans into ribbons of tattered scraps still in orbit just overhead and impossible to collect and remove. And as one blurb for a beer or a cigarette is torn to luminous illegible shreds, another will be launched to replace it.

Thus will humanity have imposed the biggest and most glaring "what if" of all. Of course, environmentalists will dispute this practice, but they will be no more effective than they are in their attempt to slow down the sale of SUVs and other gas-guzzling, energy-wasteful vehicles. The night sky has been dying for decades with the glare of light pollution from our cities and suburbs to the point where many young people have never seen the Milky Way. The explosion of space advertising will finish the job more quickly and much more thoroughly. At that time most astronomical education and research through observation will probably come to an end, or be limited to the Hubble Space Telescope, now abandoned by NASA.

To be sure, several sporadic plans to launch a glary space satellite with advertising have been put forth, but they were stopped by a group headed by former Vice President Al Gore and members of Congress in both parties. However, a determined and continued effort to launch is likely if only because the capability will soon be shared by many nations. For example, in 1993 Space Marketing Inc. proposed to launch a "Space Billboard" about one kilometer in size. Its size and brightness would rival the Moon, and more than 10,000 space debris fragments per day would be created. This and a number of other similar launches are described in the proceedings cited just below. Indeed the fiftieth anniversary of UNESCO itself was to have been celebrated by the launch of two large, tethered balloons into low orbit—fortunately, UNESCO abandoned this bit of space advertising. Space belongs to no one, yet no international treaty exists to ban these kinds of excesses and none is in sight. Recently the United Nations has become involved in a more positive way. An international symposium addressing these problems was held at the United Nations Center in Vienna in July 1999. It was sponsored by the International Astronomical Union, along with the Commission Internationale de l'Eclairage (CIE), the Committee on Space Research (COSPAR), and the Union de Radio Science Internationale (URSI), and later the proceedings were published as *IAU Symposium* 196 and formed part of the technical forum of UNISPACE III, the third United Nations conference on the Exploration and Peaceful Uses of Outer Space. A statement was prepared for presentation to the national delegations to the United Nations for their consideration and to alert them to the present and future excesses of light pollution, radio inter-ference, and space debris.

The Moon is not made of green cheese, although this old adage has been around for a long time. With men walking on an obviously rocky Moon, this bromide has lost much of its appeal, with few recalling it and fewer accepting its veracity. It points to one unfortunate possibil-ity in the future. This is the same curse that was covered above, the power of advertising. The Moon's surface, or most of it, can indeed be slathered with slogans. The cigarette, beer, or soft drink of the hour, with our technology, can be spread across the lunar disk, and once there, it cannot be easily removed. By this process the lunar surface can be spread with a patina of much higher albedo, thus brightening the Moon and the sky by several magnitudes. One by one, each layer would become a palimpsest as newer slogans were spread across older ones. Unlike the artificial satellites that in time would be rendered into shreds, the Moon's face is all but indestructible. In future, the cow will not jump over the Moon as it did in the nursery rhyme, but will jump over Coca-Cola or Joe Camel or a leading brand of soap. It is by no means too early for the International Dark-Sky Association and other environmental groups to advocate a resolution of the United Nations that will help to prohibit the use of advertising in space. The availabil-ity of nuclear weapons has acted as something of a deterrent to ex-tended warfare through good sense, among other features, but we can-not suppose this prohibition will last forever. Similarly, the pollution of space visible from below has been deterred by a similar kind of limited outrage, but again eternal vigilance will become necessary to restore and retain our dark skies.

21

THE TANGLED SKEIN
OF CELESTIAL MECHANICS

Celestial mechanics, the science of the determination and prediction of the positions of celestial objects, began with Johannes Kepler. He was the first to draw conclusions directly from observations—in his case those of Tycho Brahe. Kepler published his three laws of planetary motion, the first two in 1609 and the third ten years later. The first law, deduced from the motions of the planets, states that planets move in elliptical orbits with the Sun located at one focus. Every ellipse has two foci on opposite sides of the center and equidistant from it. All before Kepler, from Aristotle to Kepler's great contemporary, Galileo, believed in circular orbits, not from experience or the laws of physics, but because circles were considered the perfect form for curvilinear motion in a perfect universe. Kepler was the first to let the observations dictate the shape and form of orbits. With thirty years of Tycho's careful sightings of the positions of the planets, Kepler tried to fit circles until it became obvious that no circular orbit would fit the observations; this was especially true for Mars, one of the two closest planets to the Earth.

Kepler's second law of planetary motion derives from the speed of a planet in its orbit, and states that the radius vector, the name for the imaginary straight line connecting the planet to the Sun, sweeps through equal areas in equal periods of time—this allows predictions to be made about the future location of the planet in its path around the Sun just from the properties of the ellipse (see Fig. 21.1). Kepler knew that physics, not historical precedent, determined the nature of the solar system, but neither he nor anyone else of his time could then have expressed this in clear mathematical language. The third or harmonic law, published ten years later in 1619, is our familiar dictum that the square of the period of a planet is equal to the cube of its mean distance from the Sun.

The complete expression of the harmonic law was left for Sir Isaac

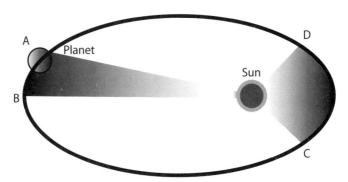

Fig. 21.1 Kepler's second law shows that if a planet takes as long to orbit from A to B as it does from C to D, then the two shaded areas will be equal.

Newton. Through his invention of calculus and other mathematics, he proved that all objects move under the gravitation of an inverse-square force field; that is, the gravitational influence varies as the inverse square of the distance between the centers of two bodies such as a planet and the Sun. Newton, pretty much by himself, developed the physics of objects moving through space. The underlying principles covering two objects mutually orbiting each other are clear and directly calculable. But whenever a third body of significant mass is considered, the calculations cannot be precisely made, and solutions for the past and future positions of any three or more of them can only be approximated. Celestial mechanics at this point quickly devolved into a huge mass of calculations, often unmanageable and all but untenable without modern calculation techniques afforded by computers.

The development of computers in the mid-twentieth century revived a stagnating field, one with many problems virtually unsolvable using calculations made by hand. Here it is important to realize that I have not made the laborious computations to know exactly the changes in the motions of the Earth if the Moon were replaced by a mass much larger than itself. The Moon's motion is by far the most complex of any object anywhere. This is because it is so close to us, and every little perturbation is observed, so in order to obtain an ephemeris of its exact position at any time, we must account for each tiny wiggle imposed by one or another object. As a consequence, the precise behavior of the Moon or the Earth involves an enormous calculative effort. With a much more massive planet nearby, the motions of the double planet would be even more prolix. That is beyond the scope and resources of this narrative. We can say that, in general, the behavior of the Earth and other planets under the alterations considered in this book are plausible within the uncertainties accepted here. Keeping this stricture in mind, we can proceed with perhaps the most significant "what if" in this story.

22

A SECOND
CHANCE

The landscape was still with only the wind breaking the stony silence. The lake lay quiet and blue, not a restful blue but intense, of a piece with the surrounding desert, gray-white and rocky, not sandy such as Arab horsemen might ride over in Saharan melodramas like Sigmund Romberg's Desert Song. Directly in front stood a stone house, now clearly abandoned. Any vegetation of the past was long gone; neither root nor branch nor twig passed for evidence of a former vertiginous existence.

The scene reminded me of Lake Mono in central California, not far north of Death Valley and immersed in barrenness as I saw it in the summer of the year 2000. But this was not the desiccated Lake Mono; no, this was Lake Mendota, the larger of the two lakes that made an isthmus of the lovely city of Madison, Wisconsin, with its handsome state capitol building dominating every view of the city. This is a painting, a masterwork of imagination composed in the mid-twenty-first century showing the now inevitable landscape that will be ours in another century.

What had happened in the intervening years of the first half of the twenty-first century? What will have brought a summer garden spot and winter fairyland to this sorry state? Why is the place certain to be utterly abandoned, with only the capitol and a handful of other ruins standing off to one side? The great dome, fourth tallest in the country, was now collapsed into itself and a devastating rejoinder to the city I once knew so well.

What had happened was simply global warming run amok. Our society had many warnings, but we couldn't bother to surrender our gas-guzzling SUVs and pickups. The petroleum industry was permitted by one sympathetic administration after another to allow the outpouring of carbon dioxide, methane, and other greenhouse gases to pollute and heat the atmosphere. By 2050 we had managed to create an inevitable second Venus in our solar system, even as that hostile planet shone as the evening star in our twilit sky. Venus was long known as a dreadful example of a runaway *greenhouse environment where the buildup of carbon dioxide began to feed on itself and heat the planet far beyond what the Sun could do by itself.*

Now the Earth is surging up in temperature just as Venus had done from natural causes millions of years in the past. We had passed the point of no return on our way to a similar hell. No more do we have the bountiful English countryside of meadows and lanes that figured in The Wind in the Willows. *To paraphrase a recent political commentary, a child asked his grandfather what he had done back fifty years ago to stop this trashing of the environment to form the desert that will come to surround us all. The old man replied that he had fought long and hard and successfully to assure that the words "under God" had remained in the Pledge of Allegiance to the Flag.*

Venus is a globe obscured by thick clouds with scarcely a break between them. It has been a mysterious world until recently, concealing its arcane secrets under this impenetrable veil. The featureless overcast atmosphere made knowledge of anything as basic as its period of rotation an impossibility; we had no idea of its length of day until radar and other modern instrumentation covertly measured it and found that it rotates but once in 244 days and backward at that. This makes the day on Venus longer by far than that for any other object known. The planet of love and beauty has finally allowed us to peek at her hidden self to find out what an inhospitable hostess she has turned out to be.

One of the most bizarre theories of the solar system was proffered by Immanuel Velikovsky, a Russian immigrant psychologist whose first book, *Worlds in Collision,* was published in 1950 by Macmillan, a very reputable publishing house. Velikovsky's principal construct called for a large chunk of Jupiter to be wrenched out of that planet by forces unknown, after which it passed close to Mars and the Earth, causing all sorts of mayhem along the way. During this promenade, the Jovian chunk, on its way to settling down as Venus, managed to stop the rotation of the Earth for a while, if only for Joshua to have his prolonged day—and all of this bedlam occurred in historical times, sometime around 1500 B.C. No mechanism is known or postulated that could possibly explain the force necessary to launch a Venus-sized mass from the body of Jupiter or to stop the rotation of the Earth and start it up again with exactly the rotation it had previously.

Some scientists made a great mistake by pressuring Macmillan not to publish the book and helped to create as a result a following for

Velikovsky among some creationists, postmoderns, and other scientific dilettantes. He went on to write and publish several more books on more or less the same theme and became quite widely known. Some years later astronomers made a much more sensible move in countering his grotesque theories and invited him to participate in a topical session dealing with planetary formation at a meeting of the prestigious American Association for the Advancement of Science (AAAS). Carl Sagan and others refuted Velikovsky's theories but the man himself did not show.

If Velikovsky can play a giant game of celestial billiards in historical times, we can surely propose that the planet Venus formed very early in the history of the solar system, not in its orbit seven-tenths of the distance from the Sun to the Earth, but in orbit about the Earth, specifically the Moon's orbit about 240,000 miles away, replacing the Moon with itself.

As was mentioned in Chapter 5, Venus more than any other planet resembles the Earth in the large overall properties—size, mass, gravitation, and average density. There the similarity ends; at the surface this celestial beauty is nothing like the Earth, but if placed in the lunar orbit at our full astronomical unit from the Sun for whatever reason, it is likely to have formed a surface and atmosphere not unlike our own. With $3\frac{1}{2}$ times the angular diameter of the Moon and an albedo some 4 to 5 times that of our airless satellite, Venus would dominate the night sky. Its brilliance at or near the full phase would be well over 40 times or four magnitudes brighter than the Moon's, and in the days around the full phase we would have very little night. We would be subject to many other alterations if our companion were Venus instead of the Moon. The two worlds could only be described as a double planet. Venus is only about 10 percent smaller in diameter than our Earth and has 81 percent of our mass; a 180-pound person standing on the surface of our world would weigh about 158 pounds on Venus.

The most profound alteration, and a near fatal one for continued life on Earth, would stem from the planet's having 65 times the mass of the Moon and hence 65 times the lunar tidal force. The ocean tides would commonly be hundreds of feet higher at high tide in many ports than they would be 6 hours earlier or later at low tide. Ports such as New Orleans and Amsterdam, which lie at or even just below sea level, could not be settled because the dikes necessary to keep out the brine would need to be enormously tall. Most of the coastal regions of the world would be uninhabitable—the tides would rise and recede so rapidly that they could not be outrun; in many cases they would drown anyone near the shore at low tide. The many great cities along our coasts could not be settled, at least not there. Tidal waves generated by the tides could regularly penetrate inland for many miles. Even river tides and bores, now scarcely a problem, could swamp floodplains with salt water and force agriculture and freshwater wildlife to abandon these areas. The Great Lakes would be subject to tides of several feet, instead of the inch or two they experience now. The tidal pull affects not only the liquid portions but also the solid, rigid Earth itself. A far greater strain on our seismic areas could stress them to the point of earthquakes and volcanism of far greater frequency and severity than is the case with the Moon.

Perhaps we might make one adjustment to this proposed double-planet system to accommodate a lessened tidal threat to human settlement. We shall therefore move Venus to twice the radius of the orbit of the Moon, thus placing the planet at a distance of about half a million miles. The effect of the move is to reduce the tides by a factor of about eight, leaving us to deal with an eightfold tidal increase, itself a giant but perhaps a just manageable impediment to the mastery of the seas.

Ours becomes a world less settled and colonized than the one we now know, for sailing from one port to another would be nearly impossible on most occasions. Not until air travel was developed would we be able to cross the oceans to colonize the Americas and Australia, the continents not connected to the Eurasian landmass. Coastal areas everywhere are unlikely to be thickly settled in such uninhabitable conditions. The use of navies is highly restricted in this case, and sea power is far behind that controlled by land armies. A nation like Great Britain would not have the armed might it has enjoyed, and its impact on world history would be much less than one occupying the heartland of the continent of Eurasia. Whenever and however the Americas were settled by Eurasian powers, China and Russia might have been at their forefront, not England or Spain.

In its new terrene home, Venus happens to rotate with a period of 33 hours and 10 minutes, at a speed somewhat slower than ours, giving it a longer day. The reason for this is probably that the Earth's greater mass has more of a tidal effect in slowing down its rotation than Venus does in slowing ours. (Similarly, the comparatively massive Venus has slowed the rotation of the Earth from 18 hours almost a billion years ago, past the 24 hours imposed by the Moon, to a longer day, but here we shall stay with our present day for convenience.) Very early in their lifetimes both planets spun around in only about 8 hours, and both have continued to slow down ever since. With its longer day, the calendar of Venus contains only 262 days since its year is the same length as ours. On average we would expect the winds on Venus to be lower than they are here due to its slower rotation. This is the result of the Coriolis effect, which brings about the deviation of a missile or of circulating air due to the rotation of the planet. With a slower rate of spin, the Coriolis effect is smaller there. Vortices, highs and lows, hurricanes, and extratropical storms are all likely to be of lower average intensity there for this reason.

Our month would be shorter, as a few minutes with a calculator can reveal. Recall that Newton's generalization of Kepler's third law of planetary motion states that $P^2 = a^3$ divided by the sum of the two masses. In this case the combined mass of the Earth-Venus system is 1.8 times that of the Earth together with the Moon. With the mean distance fixed at the same level, the length of the revolution of the two around a center of mass nearly halfway between them is 20 days, and the synodic period or lunation, the time between successive repetitions of the same phase, is about 22 days. With the stipulation that Venus is twice the Moon's distance in our drama, the month grows from 22 days to not quite 57 days, almost 2 full months, allowing but 6 of those months in the course of a year. Even at this greater distance, Venus would appear almost twice the Moon and Sun in apparent angular size. Total solar eclipses would be more frequent, but total lunar eclipses would disappear altogether since Venus is too large to fit inside the Earth's shadow.

Finally, we must make mention of the singular anomaly presented by our atmosphere and its near absence of carbon dioxide. The atmospheres of both the real Venus and Mars, disparate as they are from each other in surface pressure and temperature, are dominated by carbon dioxide; 95 percent of the atmospheres of both planets are composed of this gas. It is our world that is the exception. Some 4 billion years ago, not long after its formation, the Earth, too, had an atmosphere composed mainly of carbon dioxide. At that time Earth was not suitable for human existence, as oxygen was scarce or nonexistent. Then the water vapor, which was also abundant in the air, condensed and fell as rain, forming the oceans. On the real Venus, conditions were so hot that water remained in gaseous form, whereas Mars was so cold that most of the water remained frozen. Here in our orbit, nearly all of our carbon dioxide (all but 1 part in 20,000) came to be dissolved in the ocean waters, forming a very weak carbonic acid. Later the acid reacted with sediment on the ocean floor to form sedimentary rock rich in the carbon-ates. When plant life evolved, it absorbed the majority of the remaining carbon dioxide and expelled a plentiful supply of oxygen into the at-mosphere, creating our present gaseous envelope, which is unique in the solar system. If Venus had formed as a companion to the Earth at our distance from the Sun, it too would have had an early atmosphere not unlike the Earth's and later would also have become conducive to life. Being almost as large, it too would have produced air rich in nitro-gen and oxygen, and life would have formed there as it did here, but we must keep in mind the fact that the life forms there would almost cer-tainly differ markedly from life here, the variegations in evolutionary processes being as capricious as we now know them to be.

We now realize that evolution here as it has occurred is and has been riddled by chance events. Only a tiny shift in conditions would alter greatly the kinds of species we have. We know for instance that 65 mil-lion years ago in the fifth and last calamitous mass extinction in the his-tory of life on this planet, most species went extinct in a very short time, including all of the dinosaurs. An asteroid or comet, perhaps as much as 10 miles in diameter, collided with the Earth, making a huge crater many times its own size and spreading shock waves, fires, and a nuclear winter condition around the globe. Its devastation may have been aided by excessive volcanic activity in that unfortunate time. Had this not happened, the dinosaurs might well have survived and pre-vented the rise of mammals of size, including ourselves and the apes. Venus would have had its own set of mass extinctions unrelated to ours, with many dissimilar species forming and becoming extinct. It may well have retained the lush Mesozoic jungle habitat conducive to the dinosaurs. For all of these reasons it is nearly certain that the plants and animals of our twin would be very different from those inhabiting our own world, although DNA and the other building blocks of life there might well resemble our own.

We look up at this twin world and we see a familiar planet with oceans and continents broadly similar to ours, though differently arranged. With a surface temperature and climatic conditions similar to those here, it too appears about half covered by clouds, with white areas of ice and snow surrounding its north and south polar regions. With our telescopes we have scanned its entire surface to a very high degree of resolution. We have confirmed spectroscopically that its brownish re-gions are deserts and lie mostly between 20 and 40 degrees north and south latitude, much as do the deserts here. Greenish areas surround its equator and the midlatitudes, again mimicking our planetary climate regime. With more continental mass along the equator, the area given over to tropical rain forests is larger than ours, and the oxygen content reflects this fact, being 32 percent of the whole (with nitrogen, the dominant gas, comprising almost all the remaining 67 percent) as against our atmosphere, consisting of 21 percent oxygen and 78 percent nitrogen. The air pressure at its surface is slightly lower, but it seems as if we could breathe on Venus with little need for respiratory assistance. Land on the Earth accounts for about three-tenths of its total surface or 56 million square miles, whereas Venus favors land to a greater degree with 45 percent of its surface covered by land—this gives it 76 million square miles of land, rather more than the Earth's. The excess could simply be the result of the planet's being in the midst of an ice age, whereupon more water is frozen in its polar regions and the oceans are hundreds of feet lower than they are during interglacial periods. These kinds of conditions vary; in the extreme, the Earth had no ice at all on its surface (for the last time, incidentally, around 50 or 60 million years ago) and a great excess of it 35 million years ago, for reasons we cannot completely assess.

Our recognition of living forms on the twin world testifies in favor of a very significant discovery. On our own planet life flourished more than 3.5 billion years ago, only a billion years or less after the Earth itself was formed and very soon after the planet was cool enough for life to survive. Living organisms remained in the single-cell form as bacteria or algae until only some 550–600 million years past, when the Cambrian explosion produced many families of the larger forms, including the vertebrates. The green patches on the disk of Venus strongly imply, if not prove, that that world also passed through these two great stages —the formation of life and much later the creation of its higher forms. These parallel developments give rise to the (yet unproven) concept of life arising whenever and wherever possible.

It is abundantly clear that the landmass on Venus is bunched mostly into a single supercontinent, with smaller islands here and there. This is reminiscent of conditions on the Earth around 250 million years ago, when Gondwanaland, a southerly mass comprised of the future Africa, South America, Australia, and Antarctica rammed into the northerly mass of Laurasia, later to become Eurasia and North America. Our huge supercontinent is known as Pangaea, whose entire central region came to suffer the extremes of continental climatic regimes. Pangaea's great size may have played a role in the most lethal and severe mass extinction of all time. In perhaps 100,000 years or less, surely no more than a few million years, more than 90 percent of all living species went extinct in what must have been a huge central desert. The result was to bring an abrupt end to the era we call Paleozoic, and to leave both the land and the sea underpopulated, with many ecological niches that begged to be filled through evolution with new creatures belonging mostly to new species.

That new world ushered in the Mesozoic Era, replete with dinosaurs of every size and type, beginning with the Triassic Period, the first of three that make up the Mesozoic, the subsequent ones being the Jurassic and the Cretaceous Periods. The calamity that brought the Cretaceous to a sudden termination all those millions of years ago was the result primarily of the well-known asteroid or comet that slammed into the Gulf of Mexico at the edge of the Yucatan Peninsula, discussed in detail in Chapter 10. With an impact velocity of as much as 20 miles per *second,* it exploded with far more force than would the world's entire nuclear arsenal detonated as one, and gouged out a large portion of the planet's crust and mantle. The blast, the giant tsunamis that inevitably follow, and the nuclear winter formed by the blanket of rock and soil in the atmosphere blocking most sunlight might also finish off most existing species. Just as the Mesozoic Era died that day, a recurrence could again allow an empty planet to fill itself in with new species.

Unlike the apocalyptic K/T devastation, the P/T event seems not to have been caused by a sudden fiery spectacle suitable for Hollywood films of stomach-churning violence, but a slower process that was no less lethal in its final accounting. Although other forces may have been involved, the climatic degradation brought about by the formation of the Pangaean supercontinent was sufficiently advanced to dry up many lakes, marshes, and other sources of drinking water needed by the many land animals, and to possibly lead to a consequent enrichment of atmospheric carbon dioxide that killed off most marine life as well.

The Mesozoic Era is thus bounded in time by two giant crescendos of violence like two bookends that brought about its birth and its death. During most of the intervening 180 million years, dinosaurs, big and small, herbivorous and carnivorous, predators and scavengers, ruled the Earth.

Might that epoch surrounding the P/T boundary be like the present situation on our nearby sister world? The center of its huge landmass appears desertlike with the same Mars-red rocks that signaled the beginning of our own Mesozoic Era. Might the plant life, which we know to exist from the chlorophyll seen by our many spectrographic observations, be at a relative minimum? Pangaea was relatively free of mountain chains, especially in its vast heartland, and so appears Cytheria, our name for the great continent on Venus, taken from one of our adjectival words for the goddess of love and beauty. Many paleontologists speculate that now would be a fruitful time to colonize Venus, as the effect of *Homo sapiens* leading a parade of large mammals onto its surface might preclude a long time span dominated by dinosaurs. This assumes a feature we know to be very doubtful, namely the replication of similar species due to the repetition of similar environments, chance and chaos being the dominant determinants in any such scenario. Despite all this, our science fiction, if not our science, is ruled by a saurian Venus.

We think we know how our globe appears from space. It is certainly a mix of clouds provoked by fronts and vortices backed by the blue-tinted oceans and the land areas both green and brown, with white at the poles denoting the permanent ice caps in the Arctic and Antarctic. By day, the sunlit hemisphere would reveal little evidence of the presence of our species, but by night the dark hemisphere would be ablaze with the lights of cities and towns, the slash-and-burn fires in the Amazon basin and elsewhere, and the oil fires around the Middle East. This is the visible evidence for the next great mass extinction now well underway and caused by the global warming due to the emission of greenhouse gases from manmade sources. Species around the globe are passing into oblivion at a very rapid rate, sufficient soon to qualify as another great mass extinction, this one due to the industrial and other air pollution brought about by the dominant species on the planet.

Venus shows us evidence that it is a seismically active planet with volcanoes as active as Mount Etna, and geologic faults such as a giant cicatrice across Cytheria not unlike the Great Glen coursing over sixty miles through Scotland between Fort William and Inverness and containing Loch Ness along the way.

We have, of course, looked for evidence of cities or villages on Venus or even fields indicating a modicum of agriculture, but we have found no evidence for it. If, by any chance, intelligent creatures inhabit that world they are at best 10,000 years behind us in technological advancement, and if no high level of intelligence is present, we can turn the clock back to at least 100,000 or 200,000 years earlier. It was near that time when the brainy *Homo sapiens* replaced the comparatively intelligence-challenged *Homo erectus,* and about 10,000 years ago that our species discovered the harvesting of wheat and other crops and the domestication of animals, as an alternative to the nomadic hunter-gatherer form of existence, thus allowing stable communities to begin to form. These are perhaps the two most important transitions in our past, and neither seems to have yet taken place on Venus.

This has not thwarted a literature in realms of fantasy and science fiction rife with speculation on civilizations there technically equivalent to ours, though frequently with different life styles and motivations. Most portray societies with a mindset toward conquest, enslaving or eliminating our species and settling the Earth with their own kind. This is a variation on the theme of intelligent Martians as inherently evil creatures bent on our replacement, as in the musings of Percival Lowell and H. G. Wells, discussed in Chapter 5.

Much of our literature is of the genre that creates imaginary civilizations and even whole new intelligent species. From John Bunyan through J.R.R. Tolkien, writers have peopled new worlds of fancy; how many of these authors might have used Venus as a very convenient locale for their works. The locations of *Pilgrim's Progress,* Middle Earth, and possibly even Wonderland and Oz might have been transferred to the deliciously brilliant globe hanging just above our heads. Sir Patrick Moore, in his *New Guide to the Moon,* has much to relate about peoples' dreams and folk tales dealing with lunar inhabitants. He notes that Nevil Maskelyne, the Astronomer Royal of the late eighteenth century, denied the prospect of moon-men to even the likes of Sir William Herschel, although later Herschel agreed that the evidence for moon-men was minimal. Over the next few decades, considerable evidence showed the Moon to be as we now know it—airless and totally arid with immense and lethal temperature swings. Still the great Lunar Hoax seized the attention of a gullible world, a prank that Moore likens to Piltdown Man in the domain of immaculate deceptions. This was triggered by the location in the Cape of Good Hope of Sir John Herschel, William's astronomer son, in 1833 with a large telescope. This circumstance gave one disingenuous news reporter, Richard Locke of the *New York Sun,* the idea that he could fabricate the hoax by publishing a series of reports from Herschel that he had spotted fanciful humanoid creatures and other animals disporting on lovely sylvan landscapes within some of the craters on the lunar surface. These myths were thoroughly debunked only some time later when the stories got back to Herschel and others. Think of the abundant possibilities the sometimes verdant Venus would have given rise to—we would have had a new and fertile genre of imagined legends that would be much harder to debunk.

The inability of our earlier mariners to surmount the tides and expedite the settlement of some of our continents until recently makes for a modestly retarded level of scientific progress and as a result no space program yet exists. We have rockets that can soar to the edge of our atmosphere and can anticipate that before many more years have passed we will place an artificial satellite in orbit. Plans have been put forth endlessly regarding the prospect of exploring and colonizing Venus once the capability of space travel is achieved, but it is as yet a forbidden world glittering aloft so enticingly.

The equator of our companion is inclined only a few degrees from ours, though its tilt is less, about 18 degrees from the ecliptic. We do not expect to find the seasonal and climatological extremes of summer heat and winter cold there that we experience here.

The geocentric parallax is a term akin to the heliocentric parallax except it uses the Earth's diameter as a baseline. Parallax in this form is much too tiny to detect the distance to a star, but it is very useful for the determination of distances within the solar system. If for example we can observe a shift in the position of a nearby planet or asteroid among the background of stars from the evening after sunset to the morning before dawn, we can, knowing the size of the Earth, triangulate the distance just as we do to a star from our orbital motion.

The Moon having only one part in eighty-one of our mass means, as Newton's laws predict, that the center of gravity or barycenter will be only one part in eighty-two of the roughly 240,000-mile lunar distance from the center of the Earth. This amounts to about 3,000 miles or 3/4 the radius of the Earth; thus the barycenter is always about 1,000 miles below the surface of the Earth at the point that sees the Moon overhead at the zenith at that instant. As shown above, Venus, on the other hand, contains $65/(81 + 65) = 65/146$ of the total mass, and the barycenter must lie by that ratio along the line connecting our center to its own, or almost halfway between the two. With no space program, we cannot precisely determine the mass of Venus (or the Moon) except crudely from its size and probable average density. But knowing the motion a nearby asteroid makes every day, reflecting our geocentric parallax, an additional superimposed monthly wiggle in its motion due to the

Earth's motion in its orbit about this local barycenter can be compared in amplitude to the daily motion, and from this the mass of Venus works out very nicely.

We still await the first space vehicle to go into orbit around Venus. Look again at Newton's equation. With the mass of the spacecraft so small as to be totally negligible, Newton's generalization of Kepler's third law of planetary motion states that $P^2 = a^3/M$ where M is the mass of Venus alone since the mass of the spacecraft is negligibly small. With the distance and length of the month known with extremely high precision, the mass of the planet can be found almost exactly. In real life we first derived the mass of the Moon in just this way.

Tantalizingly close to us and in our thoughts every day, Venus has played upon human imagination in every possible manner; we are fascinated with it. People relate to Venus in a seemingly infinite number of ways. Some religious folks despair that Jesus may not yet have visited that world or converted any people who might happen to live there. They have extensive plans to colonize the new world and baptize and convert anyone there with a modicum of intelligence. Some religious groups on the fringe have taken to imagining these creatures of Venus as lustful, licentious, and minatory in other ways. If Jesus and Muhammad have not visited those beings, they are with sin and there is no telling what mendacity they may wreak if and when they arrive here. Some would have us arrive there beforehand at whatever cost in order to defuse their evil intentions.

Others seek only to plant a flag in its soil in order to create the nation state of New Russia, Nouveau France, or new states in our own great republic. Some look upon it fondly as a source of metals, oil, and other valuable natural resources, while still others see it as a giant breadbasket larger even than central North America, supplying food enough for our exploding population. Most of humanity sees the Earth's twin as something for us to use, now that we have desecrated this world so thoroughly.

Must we make a polluted mess of a second precious planet? Must there be large cities with endless slums and garish streetlights shining up to further illuminate the whole solar system? Can we not constrain ourselves and plan carefully for all members of our race so that none gain an unfair advantage? The time is soon upon us when decisions as to its use and abuse must be made. Can we avoid the psychopathic paranoids who use their considerable wiles to enchant and then oppress the masses into assisting them in the creation of tyrannies that resort to genocide and other blasphemies—do we need new Hells established on Venus? And what about the three monotheistic religions that use everyone from Abraham to Muhammad to prove that their religion has the God-given right to stifle and kill the nonbelievers?

Alas, we have no Venus, no second chance, to build a world free of the worst of human foibles; if we did we would probably see each subsegment of human society building spaceships of their own and for their own. On such a Venus we need no subsets of human culture ready and eager to find another subset to disparage and even kill off. Can Protestants, Catholics, Jews, Muslims, Caucasians, Blacks, Latinos, Americans, Russians, Arabs, Chinese, Indians, Pakistanis, fundamentalists, atheists, creationists, evolutionists, men, and women ever accept each other as equals with equal rights? Or ever agree that each of us adopt his or her own set of beliefs about supreme beings, and keep these feelings to oneself? Proselytization is just another word for coercion; let's leave it behind this time around. It belongs on the same dung heap of history as slavery and apartheid, and tribalism in general.

I look forward to a society in which the gullible play no role in the determination of science and nonscience, in which the zealots of any stripe stay behind here at home, and in which bias against those different from ourselves does not flourish. Let our next society make war on our real enemies—mosquitoes, harmful bacteria and viruses, cancer, AIDS, and other maladies.

We do in fact have that world before us; it is the Earth, not yet beyond recall. We need no second chance—if we can only act as a single species and work together.

23

CHICXULUB,
THE WORST SKY OF ALL

Earlier today we felt the ground shake, a great shudder of such violence that trees swayed back and forth and a few lost limbs or even fell over. Then nothing—all was quiet. But moments afterward great fireballs swooped across the sky, so bright that they could not only be seen in full daylight, but were even bright enough to cast shadows almost as intense as those cast by the Sun.

It did not take us long to determine the cause of this display of fireworks; an asteroid about 10 kilometers (6 miles) in diameter had just struck the Earth. Although it did so on the far side, its brutal aftermath would be experienced here and around the world.

Below the approaching missile the sight would have been quick but astonishing. Some few minutes before impact the interloper could have been seen as a swiftly growing and brightening spectacle by those directly below. On a clear day eyes would look up other than by chance only seconds beforehand, for the thing is descending at a rate of 15 miles per *second*, or nearly a thousand miles per minute. What chance would one have to seek shelter or other protection from the impending collision? If the sky were overcast, only a half-second or less before the disaster would the object be noticed. The best chance for warning would occur if it came on a clear night; then we might have a full minute to take action. Ten minutes before impact it would appear as a point of light as bright as Jupiter or Sirius, depending on the angle between its location in the sky and the Sun, hence on its phase. It would in the succeeding interval brighten rapidly flaring into a small disk a few minutes before the catastrophe, and one minute before it would appear as big and as bright as the Moon.

The sky during and just after the formation of this lethal crater must have been the most terrible sky of all time. As long as humans act together and quickly, there may never be a repetition of it. The kinds of objects responsible for calamities of this kind are divisible into two major groups, the comets and the asteroids. The former class originates beyond the Kuiper belt in the much more distant Oort cloud, and contains mostly water ice in which rocky material may be embedded, whereas most of the asteroids crowd into the space between Mars and Jupiter and are of a stony or iron-nickel constitution. Asteroid is a loose term for any minor planet that orbits the Sun, whereas the smaller among them are called meteoroids while in space. At the time of fall through the atmosphere, shining in the sky, they are meteors, and should they reach the ground, they become meteorites. The same object before, during, and after its fall can be known by all three names. Now it has become possible to spot objects as small as 100 meters in diameter or even less while they are thousands of miles off.

An asteroid with a diameter of one kilometer is estimated to strike the Earth's surface about once every million years on average. One 10 times this size, 10 kilometers in diameter, occurs every 100 million years or longer—this is the approximate size of the object that impacted the edge of the Yucatan Peninsula in Mexico and wiped out the dinosaurs 65 million years ago. Smaller objects fall more frequently; thus, a 100-meter (328-foot) object can be expected each millennium, and a 10-meter stone nearly annually. Two of the most famous falls are the asteroid or comet that struck the Tunguska region of central Siberia on June 30, 1908, and the iron asteroid that fell about 50,000 years ago and created the most famous crater, the Barringer crater near Winslow in northern Arizona. Estimates of the original sizes and masses of these objects vary; the Barringer meteoroid or small asteroid was iron-stony, perhaps 50 meters in diameter, and weighed over a million tons. The Tunguska impact was more likely of a cometary origin and composed of ice and rock of about the same size but with less mass than the incoming object in Arizona. A meteor as big as a boxcar weighing about 1,000 tons passed into our atmosphere in August 1972. It was seen glowing by thousands of tourists in the western United States and Canada in the clear daytime sky and then passed back out into space with a very different orbit but not much the worse for wear.

These larger objects are extremely dangerous and capable of extensive local or regional damage. Recently it was calculated that a one-kilometer stone could raise waves as tsunamis that might start out at a height of 2,500 feet and attain a speed of 380 miles per hour. The tsunamis could circle the globe and level coastal cities with ease. It is easy to imagine that a much larger object could bring about a mass extinction, such as the K/T event that killed off the dinosaurs and all members of over half the species on land or in the sea. Most smaller stones create minimal damage for the reason that 70 percent of the surface is ocean and much of the land is in places like Greenland, Antarctica, northern Siberia, or central Australia, all places with little or no human habitation.

A thrilling and certainly chilling account of the K/T disaster is given by Walter Alvarez in his book *T. Rex and the Crater of Doom,* an apt title for one of the disaster movies made of such an episode (perhaps starring Harrison Ford). Alvarez and his Nobel laureate physicist father, Luis Alvarez, and two other scientists coined the theory that described this cataclysm. They raised once again, and this time decisively, the debate as to whether nature changes only very slowly in a process known as uniformitarianism, or can change suddenly through catastrophism. During the nineteenth century, with the rise of geology and the influence of Charles Lyell and other geologists and Charles Darwin of evolutionary fame and his colleagues, the uniform view became accepted by most scientists to the point of dogma. Throughout the twentieth century, this view slowly eroded and widespread catastrophes were considered. The Alvarez theory was debated for years after its initial publication in 1980 until the site of the event, the "smoking gun," was discovered and recognized in 1991 in the form of a crater more than 100 miles in diameter located partly on the Yucatan Peninsula of Mexico and partly underneath the offshore waters of the Gulf of Mexico. Now known as the Chicxulub crater, it is named for Chicxulub Puerto, a coastal village near the town of Progreso, a port city near Mérida, the capital and largest city in the Mexican state of Yucatan (see Fig. 23.1). Nothing of the crater is visible as it lies under layers of sediment deposited after the event, but the town of Chicxulub—a Mayan word pronounced "cheek-shoe-lube," according to Alvarez—has achieved a modicum of fame or notoriety as a result.

When a rock 6 to 10 miles across hits our planet at some 20 miles a second, 150 times as fast as jet airliners fly, bad things happen. Lately we have had a taste of the damage that impacts at high speed can achieve when in July 1994 comet Shoemaker-Levy 9 hurtled into Jupiter. The comet had earlier been captured by the big planet and was in orbit about it. The huge Jovian gravitation tore the comet into about twenty-one pieces, each a mile or more in size. One by one like soldiers marching to their doom the fragments tore into the Jovian atmosphere and left visible holes in it. Jupiter gets struck in this fashion much more often than any other world and helps clean up the detritus still in the

Fig. 23.1 View of the seacoast of the Mexican village on the Yucatan Peninsula, Chixculub, located near the center of the huge crater that was formed by the asteroid or comet that collided with the Earth 65 million years ago and killed off the dinosaurs.

solar system that might otherwise strike the Earth. We had ringside seats for a calamity that we hope to prevent here.

What would we experience if an exact twin of this K/T level of devastation were to happen again today? We think it would be along these lines. The asteroid (or comet) comes into the impact site at the edge of the Yucatan Peninsula from the southeast at some 20 miles per second or about 70,000 miles per hour; it covers the 6 miles between the flying altitude of commercial jet aircraft and the ground in a single second. Bigger and more massive than Mount Everest, this bulk at this velocity causes an explosion with perhaps 10,000 times the force of the entire nuclear arsenal of the entire human race at the peak of the Cold War, or some 100 million megatons.

At this speed a shock wave is formed in and through the surrounding rock. Some rocks simply vaporize under the compression. The hole is over 100 miles in diameter, but more amazingly, it is more than 25 miles deep until the surrounding hole it has made promptly collapses into itself. For comparison, we can refer to the deepest hole in the ocean floor, a place in the mid-Pacific Ocean known as the Marianas Trench. This lowest spot on the solid surface of the Earth is 7 miles in depth below sea level; in fact, the variation around the entire globe from the summit of Mount Everest to this trench is less than 13 miles, about half the depth of this seconds-old crater. The Earth's crust, its outermost solid layer, is composed mostly of sedimentary rock of relatively light density. Below this layer lies the mantle, a much thicker zone of heavier, denser, mostly silicate rock somewhat heated by warmer stuff that forms the metallic core, consisting of iron and nickel in a partly molten liquid state. The outer crust is a basaltic layer some 3 to 5 miles thick on the ocean floor and twice this thickness on the continents, which are formed primarily of granitic types of rock. The continental stuff is largely quartz and feldspar dominated by sodium and calcium, whereas the oceanic floors consist of calcium and magnesium compounds. The comet or asteroid nemesis of the dinosaurs may have penetrated through the outer crust entirely since the region of the impact crater is a mix of more than one type of rock.

Nothing else even remotely on this scale has happened on the planet in over 200 million years; the entire human species has no direct experience with it. We can only piece together from indirect evidence just what happened on that day on which the Mesozoic Era died.

Seconds after the impact comes, the shock wave slams out of the great hole in the Earth and momentarily compresses the rock, which then decompresses and even partly vaporizes on the spot. This is a sonic boom like no other, blowing out eardrums for miles in every direction. The atmosphere heats up to tens of thousands of degrees on any scale, many times as hot as the surface of the Sun. Ejecta come flying out in the form of heated, melting fragments of rock. One and then another of these fireballs strikes the ground somewhere nearby in every direction.

Cities throughout the Americas would die on the spot, as would most of their inhabitants. The ensuing fireball would set forests and crops ablaze; then would come tsunamis thousands of feet high, flattening every skyscraper in coastal regions, at least. The United States would cease to exist. Civilization would come to an abrupt end, although *Homo sapiens*, numbering over 6 billion and settling every land area on Earth except Antarctica, would survive in small groups, which might have to revert back to a hunter-gatherer type of existence, much as they did during the last ice age. Now all of humanity faces an even greater threat; dust, particulate matter, and acid rain come pouring into and, months later, out of the atmosphere. We pass through a nuclear winter far more extensive than any nuclear exchange, no matter how extensive. The junk in the air cuts off most of the incoming solar radiation for many months, and the whole planet cools off markedly. Temperatures plunge to levels that wipe out more than half of all existing species, animal and vegetable.

Following the Alvarez theory, catastrophism became an acceptable alternative to the uniformity of earlier times. The dinosaurs may have faced other stresses, but this sudden lurch was just too much for the survivors. Nature at its worst can occur suddenly or gradually; in this case of a mass extinction both were to blame.

Do we face another blast like this one? Yes, of course, sooner or later we do. Sometime another interloper will intrude where we do not want it and if large enough will promptly end all life on the planet. But in our favor we can say that asteroids or comets larger than this unholy terror are exceedingly rare. We know of almost all asteroids of this size (5 to 10 miles across) in the solar system and can track them. Comets leave us less time to act, because by the time they are discovered they are maybe as far away as Jupiter and if aimed at us will arrive in just a few years. Still, our species is not far from guaranteeing our continued existence against the one and only threat that could end the species forever.

Even much smaller missiles of around one kilometer in size strike every one million years on average. The odds are heavily in our favor, despite the close calls trumpeted by the media in recent post-Alvarez years. Now and then we hear that an asteroid will pass frighteningly close to the Earth by the year 2028 or some such. A few days after the scare another report comes along that it will miss us by a reassuringly much greater distance. Whence the original scare? It comes from the fact that the orbits of most asteroids are known but not precisely well known. When the observed coordinates of one indicate a hit or a very near miss, astronomers immediately search photographic records of the sky taken in the past at the Harvard College Observatory and elsewhere. Whenever the image of the asteroid in question shows up on an early observation, as it almost always does, the orbit can be derived to a far greater precision. The almost inevitable result is that a near miss becomes a country mile.

Think of a dartboard; if one's aim is only good enough to hit it half the time, one will score very few bull's eyes, for the cross section of a bull's eye may be well under 1 percent of the area of the entire board.

How many darts will stray by the average error the size of the radius of the board directly into the tiny bull's eye, compared to the number aimed for it that strike around the edge of the board? The cross section of the entire Earth is mighty small in the solar system, and almost all missiles will miss it cleanly.

Most of these flying rocks stay out there where they belong, well between the orbits of Mars and Jupiter. But occasionally one is discovered whose orbit crosses our own—these are the "Earth-crossers." Three of the earliest were named Apollo, Amor, and Aten, and each gave its name to a group of asteroids that share their orbital characteristics. Thus the Apollos, Amors, and Atens collectively make up the class of Earth-crossers. How many of them are there? This is a very difficult number to derive. The prominent astrogeologist Gene Shoemaker has probably studied this problem more than any other scientist. A few years ago he concluded that about 1,100 asteroids larger than one kilometer or 0.6 mile proceed around the Sun in orbits that can cross the orbit of the Earth. This is well under 1 percent of the total number in the system. Most individual members of both these groups are not named. In order to formally receive a name from the designating body, the International Astronomical Union (IAU), the orbit of an asteroid must be known to a very high level of precision—so precise that we can relocate it again with no problem. This usually means observations extending over a longer time span than the orbital period.

There may be 10 times as many of only 100 meters in diameter, but at this level a direct hit or a near miss would be necessary in order to flatten a city. Large urban areas occupy far less than 0.1 percent of the planet's surface, thus most hits of this size would do limited damage.

Comets entail a dicier situation. Comet material lies out beyond Neptune and Pluto and the other Kuiper belt objects. They stay out there, slowly moving along in their centuries-long orbits unless one of them nudges another out of its orbit into a smaller one of high eccentricity, which may bring it much closer to the Sun. If the perihelion of the new orbit is of one astronomical unit or less, or if Jupiter or one of the other major planets further perturbs it into an Earth-crossing orbit, the comet can become a problem. Comets come in all sizes with about the range in diameter of asteroids; Halley's comet, the first to be imaged by a passing space probe during its 1986 perihelion passage, is about 10 miles long and half that in width and thickness. Being about as large as the K/T object whose collision ended the Mesozoic Era, it would wreak a similar havoc were it to plow into the Earth as did its late cousin, comet Shoemaker-Levy 9, into Jupiter. Two spectacular recent comets reveal a range of size; comet Hyakutake was small, about a kilometer in size, when it whipped by us in 1996, but the following year comet Hale-Bopp lumbered along at a greater distance from the Earth at around 25 miles across, a size to end all life on the planet should it strike us.

We have a shorter warning period for a comet than for an asteroid, at least a large asteroid. We have discovered the majority of Earth-crossing minor planets and have well-defined orbits for them—if one should get out of hand, due to perturbations from a major planet or a collision with a smaller one, we will know about it, maybe well in advance. But by the time a comet is discovered, even a large one like Halley or Hale-Bopp, it is probably about to cross Jupiter's orbit 5 astronomical units from the Sun and will get to our distance in about two years—perhaps not enough time for us to take evasive action.

Evasive action means for all practical purposes deviating the object in its orbit, causing it to miss the Earth. Blasting it apart with a nuclear weapon is unwise, for a few of the largest pieces might well be aimed directly toward our planet and do nearly as much damage as the original body. It is far safer to leave it intact but nudge it into an alternate orbit; the only immediate way for this to be accomplished is through a nuclear explosion next to it with the force pushing it away, but how do we know in advance whether the intruder is a single corporate chunk of rock or ice or whether it might quite possibly be a loose collection of several pieces bound by little more than their collective mutual gravitational attraction, as some are thought to be?

Other solutions have been proposed. One possibility calls for the ejection of a bunch of very small particles, even as small as ball bearings, directly at the interloper, thus allowing each particle to collide with it and drive it infinitesimally away from its original path, via Newton's laws of motion. The sum total would, under this hypothesis, be equivalent to a single larger projectile.

In the days after the announcement by the Alvarez group of the K/T mass extinction brought about by a collision with a sizeable object, evidence of other smaller mass extinctions since that calamity were sought and thought to be found by some. For a while these extinctions appeared to happen periodically with a spacing of some 30 million years. One meeting, held in Tucson, Arizona, in January 1985, at which the author was a participant, was among those that intended to focus on the means by which periodic mass extinctions might occur and the astronomical reasons for their alleged periodicity. It may not have been recognized then as it is now, that the data for extinctions was stochastic in nature; that is, randomly spaced.

Among the conjectures involving astronomical causes of evenly spaced events was the presence of giant molecular clouds in interstellar space through which the solar system might pass from time to time, thus turning on a nuclear-winter effect from the cloudy material passing in front of the Sun, cooling the Earth to the point where some species would freeze into extinction. Another involved a periodic motion known to occur whereby the solar system passes through the plane of the Milky Way galaxy in an up-and-down motion of about the requisite period. Neither of these explanations remained satisfactory for long.

The one that generated the greatest interest and speculation required the presence of a solar companion star in a large, sprawling, highly eccentric orbit. Chapter 3 featured a rough equivalent to this star by positing Proxima, a small red dwarf star in a near circular orbit around a double star formed of the Sun and a much closer companion star. This star was even given the name of Nemesis, a minor Greek deity, the goddess of retributive justice or vengeance. She would dole out punishment for our collective sins on a regular basis, much as a possible companion may have done at Sodom and Gomorrah in biblical times.

The size and nature of this proposed stellar orbit were strictly constrained; they required a periodicity of some 30 million years with the perihelion, the closest distance between Nemesis and the Sun to be within the Oort cloud of icy trans-Neptunian objects (comets) orbiting around 10,000 astronomical units from the Sun. From Kepler's harmonic law as emended by Newton, we can calculate that the mean distance of Nemesis must be of the order of 100,000 astronomical units or about 1.5 light years. Since the perihelion distance must be relatively nearby, the aphelion, the farthest point, must be almost twice this distance or between 2.5 and 3 light years, more than half the distance to the nearest star system, Alpha Centauri.

Many problems rise to defeat this hypothesis more certainly than was first realized in the 1980s. First and foremost, the Sun, carrying the planets along with it, circles the center of the galaxy about once every 200 to 250 million years; hence our system has made about twenty circuits about the center since its formation. During that time many stars would have passed us by a distance closer than that of Alpha Centauri, and their cumulative tidal effect would have separated any such solar companion from the rest of the solar system well over a billion years ago. The much smaller and more circular orbit of Proxima is far more stable and might well have remained so over the last 4.6 billion years of our solar system. The chance of a capture by the gravitation of the Sun of an object so far away in more recent times is vanishingly small—here Ockham's Razor comes into play in that many factors would be necessary to explain the capture.

Second, for a star to remain a luminous object over several billion years it must have a mass of at least 8 percent of the Sun's mass or about 80 times the mass of Jupiter. Masses less than this cannot raise the central temperature to the point where thermonuclear fusion converting hydrogen into helium can begin. Smaller masses than this occur but they shine as stars for a much shorter time and cool off before fusion can get underway. If Nemesis had been big enough to be a proper nearby star, it would long since have been recognized as one. Our mapping of the heavens is well beyond the point where Nemesis could have been missed. Even at aphelion, its most distant point, it would appear not fainter than magnitude 9 to 10; its spectrum and high parallax motion would have been well known over the last few decades at least. Any smaller star would be a brown dwarf and have faded into darkness more than a billion years ago.

The reason for the highly eccentric orbit for this companion lies in the assumption that at perihelion, occurring once about every 30 million years, it is close enough to pass into the Oort cloud of distant comets and perturb a number of them into orbits that would bring them into the inner solar system so that one or two of them could collide with the Earth. The presence of Proxima, discussed in Chapter 3, would be very unlikely for this same reason. It would have perpetually shaken up the Oort cloud to the point that life on Earth would have been bombarded into obsolescence. Searches for Nemesis were made but no candidate for this death star was found. In recent years the mass extinctions since the K/T event have been found to be random through time, lessening any need for a periodic death star.

What would we see in the event of a K/T cataclysm happening today? The effects would be of short duration since they would be limited to our own atmosphere. After the shock waves and firestorms, we would see a nuclear winter sky, one darkened to the point where little or no sunlight permeates the air for some months. Years afterward the sky would appear unchanged by the collision. If the same event were to take place on the Venus of the preceding chapter, wherein that planet lies only twice the actual distance to the Moon, we would observe a series of blinding flashes should the object strike the hemisphere turned towards the Earth at the time. This would be followed by months of clouds prevailing over the surface, the tops of which would glisten in the sunlight just as clouds on the real Venus do now. Then the atmosphere would gradually clear and be restored to its customary state, but with many empty niches among its life forms, gradually to be filled with new genera and species evolving from the surviving ones. In this way it can be seen that the twin planets would pass through different histories, an even stronger reason for great differences in flora and fauna on the two worlds than mere failure of replication would provide.

For over a century schemes have been proposed to provide inhabitants of Venus, if any, with visual evidence of intelligent life here. Plans have included stripping parts of the northern taiga forests into triangles or other geometrical shapes and sending wireless transmissions of various kinds at a number of wavelengths. Such plans have all been dropped because any view from there of our night side would clearly reveal our overlighted cities to the naked eye, with Las Vegas the brightest of all.

PERSONAL NOTE

Three of the illustrations in this book are taken from paintings by Chesley Bonestell (1888–1986). Bonestell was the premier illustrator of imagined scenes on other worlds. Two of the three paintings, shown in Chapter 7, illustrate Saturn and its glorious system of rings, and the third, in Chapter 17, depicts a globular cluster in very close proximity to the Earth. It is a pleasure to thank Ron Miller for permission to reproduce these paintings. One of them also appears on the front cover. This is a view of Saturn as it would appear from Titan, its largest satellite. The painting is from 1946, just after G. P. Kuiper discovered that Titan possesses an atmosphere, thought to be a thin one composed primarily of methane. Recent space probes have revealed that Titan has a much thicker atmosphere, mostly of nitrogen, of such density that the sky would be perpetually cloudy, so that such a view of Saturn as this would not be possible from Titan's surface. In any event the picture is rightfully one of the most celebrated in all of astronomy and has been thought to have encouraged a number of young people into the consideration of astronomy as a career.

Without the assistance of a number of others this book would not have been possible. Once again John Wareham capably produced and reproduced all of the figures. Gabriele Zinn again dealt with the permissions needed for some of the illustrations. Sally Brady, my agent, and Audra Wolfe, my editor at Rutgers University Press, provided help and advice of necessity and merit that made the project a reality. I am also indebted to my wife, Joan, and my daughter, Amy, for reading parts of the manuscript and for helpful suggestions.

BIBLIOGRAPHY

Alvarez, Walter. *T. Rex and the Crater of Doom*. Princeton: Princeton University Press, 1997.

Asimov, Isaac. "Nightfall." In *Nightfall and Other Stories*. New York: Ballantine, 1984.

Barzun, Jacques. *From Dawn to Decadence: Five Hundred Years of Western Cultural Life, 1500 to the Present*. New York: HarperCollins, 2000.

Bonestell, Chesley, and Willy Ley. *The Conquest of Space*. New York: Viking Press, 1949.

Boorstin, Daniel. *The Discoverers*. New York: Random House, 1983.

Buchanan, Mark. *Ubiquity: Why Catastrophes Happen*. New York: Three Rivers Press, 2000.

Butterfield, Herbert. *The Origins of Modern Science*. New York: Free Press, 1997.

Chapman, Clark R., and David Morrison. *Cosmic Catastrophes*. New York: Plenum, 1989.

Comins, Neil F. *What If the Moon Didn't Exist? Voyages to Earths That Might Have Been*. New York: Harper Perennial, 1993.

Cox, Arthur N., ed. *Allen's Astrophysical Quantities*. 4th ed. New York: Springer-Verlag, 2000.

de Santillana, Giorgio. *The Crime of Galileo*. New York: Time Life, 1955.

de Santillana, Giorgio, and Hertha von Dechend. *Hamlet's Mill: An Essay on Myth and the Frame of Time*. Boston: Harvard Common Press, 1969.

Duncan, David Ewing. *Calendar*. New York: Avon Books, 1998.

Freud, Sigmund. *Moses and Monotheism*. New York: Vintage, 1955.

Gingerich, Owen. *The Eye of Heaven*. New York: American Institute of Physics, 1993.

Goldsmith, Donald, ed. *Scientists Confront Velikovsky*. Ithaca: Cornell University Press, 1983.

Gottlieb, Anthony. *The Dream of Reason: A History of Philosophy from the Greeks to the Renaissance*. New York: W. W. Norton, 2002.

Hartmann, William K. *Moons and Planets: An Introduction to Planetary Science*. Monterey, Calif.: Brooks/Cole, 1998.

Hirshfeld, Alan W. *Parallax: The Race to Measure the Cosmos*. New York: W. H. Freeman, 2001.

Hoskin, Michael. *The Cambridge Illustrated History of Astronomy*. Cambridge: Cambridge University Press, 1996.

Huntington, Samuel P. *The Clash of Civilizations and the Remaking of World Order*. New York: Simon and Schuster, 1997.

Kuhn, Thomas S. *The Copernican Revolution*. Cambridge, Mass.: Harvard University Press, 1957.

———. *The Structure of Scientific Revolutions*. 2nd ed. Chicago: University of Chicago Press, 1970.

Lang, Kenneth R. *Astrophysical Data: Planets and Stars*. New York: Springer-Verlag, 1991.

Minnaert, Marcel. *The Nature of Light and Color in the Open Air*. 1954. Retranslated and reissued as *Light and Color in the Outdoors*. New York: Springer-Verlag, 1995.

Mitton, Jacqueline. *A Concise Dictionary of Astronomy*. Oxford: Oxford University Press, 1991.

Moore, Patrick. *The Data Book of Astronomy*. Philadelphia: Institute of Physics, 2001.

———. *New Guide to the Moon*. New York: W. W. Norton, 1976.

Morrison, David, and Tobias Owen. *The Planetary System*. Reading, Mass.: Addison-Wesley, 1988.

Sobel, Dava. *Longitude: The True Story of a Lone Genius Who Solved the Greatest Scientific Problem of His Time*. New York: Walker, 1995.

Smoluchowski, Roman, John N. Bahcall, and Mildred S. Matthews, eds. *The Galaxy and the Solar System*. Tucson: University of Arizona Press, 1986.

Spengler, Oswald. *The Decline of the West*. Oxford: Oxford University Press, 1991.

Taub, Liba Chaia. *Ptolemy's Universe*. Peru, Ill.: Open Court, 1993.

Tuchman, Barbara. *A Distant Mirror: The Calamitous Fourteenth Century*. New York: Knopf, 1978.

Upgren, Arthur. *Night Has a Thousand Eyes: A Naked-Eye Guide to the Sky, Its Science and Lore*. Cambridge, Mass.: Perseus Books, 1998.

———. *The Turtle and the Stars: Observations of an Earthbound Astronomer*. New York: Henry Holt, 2002.

van Helden, Albert. *Measuring the Universe: Cosmic Dimensions from Aristarchus to Halley*. Chicago: University of Chicago Press, 1985.

Velikovsky, Immanuel. *Worlds in Collision*. Mattatuck, N.Y.: Amereon, 2000.

Zeilinga de Boer, Jelle, and Donald Sanders. *Volcanoes in Human History: The Far-Reaching Effects of Major Eruptions*. Princeton: Princeton University Press, 2001.

INDEX

INDEX

INDEX

INDEX

ABOUT THE AUTHOR

Arthur Upgren, Ph.D., is Emeritus Professor of Astronomy at Wesleyan University and Senior Research Scientist at Yale University. He has written three previous books and many research articles in his field. He is active in reducing light pollution and restoring the night sky to its pristine condition.